I0053524

CWA

Certified Wireless Analyst

Study Guide 2022

2231 Wireless Fundamentals
2232 Mobile Communications
2233 Fixed Wireless

Eric C. Coll, M.Eng.

Teracom Training Institute
www.teracomtraining.com

The publisher offers discounts on this book when ordered in quantity. For more information, or to place an order, please contact the publisher: Teracom Training Institute, Ltd.
Publishing Division
1-877-412-2700
www.teracomtraining.com

Copyright © 2022, Teracom Training Institute, Ltd.

All rights reserved. No part of this document nor the accompanying presentation may be reproduced or transmitted in any form or by any means, electronic or mechanical, including but not limited to duplication, photocopying, scanning, peer-to-peer file sharing, downloading or by any other information storage and retrieval system, without permission in writing from the copyright holder.

Trademarks and registered trademarks referenced in this text are the property of their respective holders and are used for identification purposes only. Teracom is a registered trademark of Teracom Technologies Inc.

Softcover print edition T4213 ISBN 9781894887687

2022-01 R3

Printed in Canada

Notice: The information contained in this document is provided as general background information only. Design and implementation of a communication system requires professional advice to identify and resolve issues specific to that particular system, including but not limited to performance and security issues. Additionally, while we have striven to be as accurate as possible, we make no representation of fitness or warranty that the information provided is 100% accurate. The information in this document is not to be relied upon as professional advice, nor is it to be used as the basis of a design. Users of this document agree to hold the author and Teracom Training Institute Ltd. harmless from any liability or damages. Acceptance and use of this document shall constitute indication of your agreement to these conditions.

A little learning is a dangerous thing;
Drink deep or taste not the Pierian spring:
There shallow draughts intoxicate the brain,
And drinking largely sobers us again.
                    – Alexander Pope

Preface

The Certified Wireless Analyst certification package is three courses delivering the core technical knowledge needed by anyone serious in the wireless business today - plus TCO Certification to prove it.

We're going to spend most of our time talking about mobile communications, because one of the great things about wireless is you can move around while communicating, and people are willing to pay money for that.

If you added up all of the industry associated with mobile communications: everything from selling handsets and providing customer service, all of the people who work for the carrier, the trucks they buy, people getting jobs as riggers installing equipment on towers, the insurance they have to pay for – the mobile communications industry shows up as part of the gross domestic product of every country in the world.

Course 2231: Wireless Fundamentals

We'll start off with some general principles: why we use radio instead of light for wireless communications, for example; then look at the wireless spectrum, the different radio bands that have been allocated for different services; then talk a bit about analog radio, and spend a fair amount of time talking about digital radio, which is actually the use of modems over radio and how modems work, then finish with penetration, propagation and fading.

Course 2232: Mobile Communications

The second course in the certification is mobile communications and mobility. We'll start off with basic concepts and terminology in the mobile business, the idea of cellular radio, then first generation frequency-division multiplexing and analog; then second-generation digital.

Putting the technologies aside for a moment, we'll look at PSTN phone calls, which are called "voice minutes" on a billing plan, and then how mobile Internet is provided, of course called a "data plan". Then to summarize, we'll look at Mobile Network Operators, Mobile Virtual Network Operators, and how roaming works.

Then we'll continue through the generations and technologies, starting with second-generation TDMA technologies, called GSM in most of the world, and called TDMA in North America. This is legacy technology; there are no questions about GSM or TDMA on the certification exam anymore, so the lessons on TDMA and GSM can be skipped if desired; but knowing what GSM actually was is part of the base knowledge of mobile communications. Literally billions of people had GSM service.

Two warring factions emerged with the second generation as to how the radio bands would be shared amongst the users: the TDMA/GSM people on one side, and the CDMA people on the other side.

With the third generation, all of the serious technologies were CDMA, but the two warring factions didn't stop warring; they just changed what they were arguing about. On the GSA/TDMA side there was UMTS wideband CDMA for third generation, and in the other camp was the 1X version, and both of them had data-optimized variations: HSPA and 1XEV-DO respectively.

Along the way, we will discuss the idea of spread spectrum and how that's implemented with CDMA, and some operational characteristics of CDMA.

Then we'll see how Steve Jobs ended the standards war for fourth generation with his iPhone by only allowing carriers in the GSM/TDMA/UMTS camp to have the iPhone. The carriers that were in the other camp threw in the towel and went with the GSM/TDMA/UMTS camp's plan for 4G, which was called LTE.

With LTE, we're back to frequency-division multiplexing; but of course, there are some huge advantages compared to the first generation: the ability to be assigned multiple channels and run modems on them for massive parallel downloads. We'll spend some time revisiting modems for radio, and end up understanding how LTE can transmit different groups of six bits to many different people at the same time on the same carrier.

Then 5G, which is called New Radio in the standards bodies. We'll discuss the immediate impact of 5G: more bits per second, and also the new spectrum that is becoming available for mobile wireless communications at the same time as 5G.

We'll talk about the design goals and use cases: enhanced mobile broadband (more bits per second for your cellphone), massive machine-type communications for the Internet of Things, characterized by very low bit rates, and then the thing the newspapers like to talk about when they mention 5G: the ultrabroadband millimeter-wave bands.

We'll finish off with a roundup of the spectrum sharing technologies: FDMA, TDMA, CDMA and OFDMA.

Course 2233: Fixed Wireless

In the third and final course in the certification package, we'll cover other radio systems, primarily fixed wireless, where the question of mobility doesn't come into play.

We'll start off with light: infrared, which is used in remote controls, Bluetooth which has become ubiquitous, and Wi-Fi, more properly called 802.11 Wireless LANs; and we'll talk about security on wireless LANs and WPA2-Personal.

Then we'll go over broadband internet access to the home via fixed wireless, then at the other end of the scale, Low-Power Wide-Area networks for the Internet of Things, point-to-point radio systems and satellite communications including Starlink.

Let's get started!

Contents

Course 2231

Wireless Fundamentals

1 Radio

"Wireless" generally means the use of *radio*, which is electromagnetic waves vibrating at frequencies measured in Gigahertz (GHz), that is, vibrating 10^9 or a billion times per second.

Wireless could also mean using electromagnetic energy vibrating on the order of 10^{14}, hundreds of trillions of times per second (this is called light). However, one of the problems in wireless communications is obstacles. It turns out, that the higher the frequency, the longer distance it takes for energy to *refract*, or bend around an object.

Light does refract around objects – this is how it is possible to tell that there are planets around other suns – but the length of the shadowed area behind the object is too long for use on a terrestrial scale.

1.1 Definition of "Radio"

If the frequency of the energy is reduced, the length of the shadow behind an obstacle shortens. In addition, lower-frequency energy can penetrate through objects like walls and clouds more easily (there's a reason why fog horns are very low frequency). For these reasons, wireless communication systems tend to use energy at Gigahertz frequencies, two or three hundred thousand times lower than light, and call it *radio*.

Accordingly, most of this course is on communications centered at Gigahertz frequencies, in frequency bands with widths measured in Megahertz (MHz), which is 10^6 or millions of Hertz wide.

1.2 Applications for Radio

Radio is used in many kinds of systems with different applications. This includes everything from demagogues broadcasting angry rants on talk radio shows using analog AM, to mobile cellular systems for telephone calls, web surfing, analog and digital video, *trunked radio* for police communications, fixed wireless to remote residences, short-range wireless

LANs, geosynchronous communication satellites, Low Earth Orbit satellites, and more.

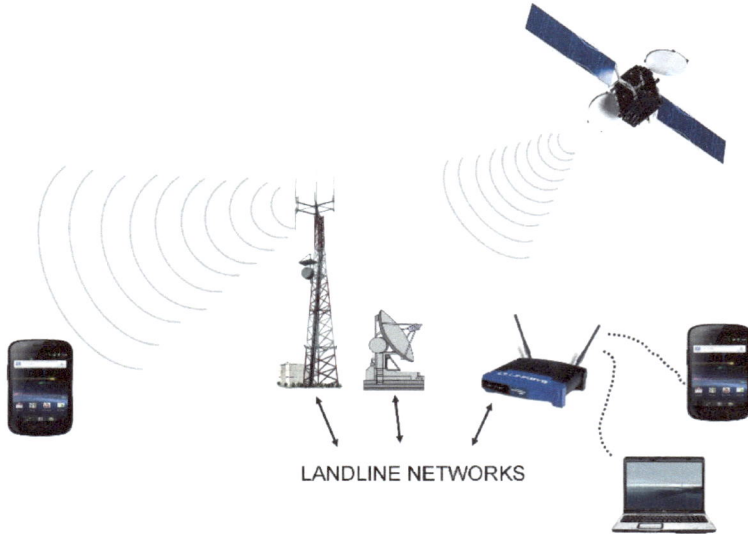

FIGURE 1 APPLICATIONS FOR RADIO

Video broadcast, two-way voice communications and point-to-point digital microwave communications were the biggest applications for radio in the past, whereas mobile voice and data communications is a significant business in the present. In the future, wireless will be ubiquitous.

1.3 Representing Information Using Radio

To represent information, a single pure frequency, called a *carrier frequency*, could be chosen, and the amplitude (volume) of the carrier frequency varied in a continuous fashion as a representation or *analog* of the strength of the sound pressure waves coming out of the speaker's mouth. This called Amplitude Modulation (AM).

Alternatively, the frequency of the carrier could be varied continuously as an analog of the strength of the sound pressure waves. This is called Frequency Modulation (FM).

Representing 1s and 0s is a more complex task. Since radio bands do not include zero Hertz, sometimes called *DC*, pulses cannot be used to represent 1s and 0s as on copper wires.

Instead, it is necessary to use techniques similar to those used in landline modems to represent the 1s and 0s. A carrier frequency is again used, but instead of varying it continuously, only specific discrete changes in the carrier frequency are defined. In the simplest example there are only two states of the carrier frequency, for example, "loud" and "quiet". One of the states is chosen to mean 0 and the other state means 1.

This was called *shift keying* in the ancient world of telegraphs: the carrier is shifted between the two states by keying a key on a telegraph transmitter, like a key on a keyboard: depressing the key produces a carrier in one of the states, releasing it producing the carrier in the other state.

Different states of the carrier can be implemented with different amplitudes, called Amplitude Shift Keying (ASK), different frequencies, called Frequency Shift Keying (FSK), two phases (relative position in time), called Phase Shift Keying (PSK) or combinations thereof, called Quadrature Amplitude Modulation (QAM).

2 Wireless Spectrum and Radio Bands

2.1 The Need for Regulation

Every country has the sovereign right to manage energy at radio frequencies in its territory. When the airspace through which the radio waves travel is public property (which is most of the time), and there is contention for its use (which is all of the time), regulation is required to allow the rational use of the shared resource.

In the United States, the Federal Communications Commission (FCC) regulates the wireless spectrum for use by other than the Federal government. Industry Canada, which includes the former Department of Communications, regulates the airwaves in Canada.

Canada, Mexico and the USA coordinate radio frequency usage due to geographic proximity. Likewise, the International Telecommunication Union (ITU) in Geneva facilitates coordination between other countries, particularly outside of North America.

2.2 Spectrum

The range of radio frequencies, called the *radio spectrum,* is divided into blocks of frequencies allocated for different services. Allocations are divided into allotments, which are bands of frequencies assigned to specific users or to the public.

A license to emit radio-frequency energy at the specified frequencies in a specified area is issued by government to record the assignment of the allotment and the conditions for its use.

← more capacity - - - better transmission →

AM Radio	0.5 – 1.7	MHz
CB Radio	26 – 27	MHz
TV Channels 2-6	54 – 88	MHz
FM Radio	88 – 108	MHz
TV Channels 7-13	174 – 216	MHz
Maritime	457	MHz
600 MHz band	614 – 698	MHz
700 MHz band	698 – 806	MHz
800 MHz Cellular NA (1G →)	824 – 891	MHz
900 MHz ISM Band NA	902 – 928	MHz
900 MHz GSM Band	880 – 960	MHz
1800 MHz GSM Band	1710 – 1880	MHz
1900 PCS NA (2G →)	1850 – 1990	MHz
4G, 5G: many between	1800 – 2400	MHz
2.4 GHz ISM Band NA Wi-Fi, Bluetooth, phones microwave ovens	2.4	GHz
2.5 GHz Mobile & Fixed	2500 – 2690	MHz
3.5 GHz Fixed Wireless	3550 – 3700	MHz
3.7 GHz C-Band for 5G	3700 – 4000	MHz
5 GHz U-NII bands NA Wi-Fi, point-to-point	5.15 – 5.925 5.925 – 7.125	GHz
Millimeter-wave 5G	28, 30, 47	GHz

FIGURE 2 WIRELESS SPECTRUM AND FREQUENCY BANDS

2.2.1 Capacity vs. Performance Tradeoff

It turns out that the capacity, determined by the widths of the bands, is very good at high frequencies. However, transmission characteristics, like penetration through walls, are not so good. Lower frequencies experience much better transmission characteristics – but have lower capacity.

The sweet spot for good capacity with long range and good in-building penetration appears to be within the 500 MHz – 2500 MHz range with current technologies.

2.3 Frequency Bands

Figure 2 lists some of the bands that are currently in use, beginning with old-fashioned analog Amplitude Modulation (AM) radio and Citizens' Band (CB) radio.

2.3.1 Broadcast Television

Allocations for television broadcast, also called Over the Air (OTA), were organized as frequency channels, each notionally 6 MHz wide.

Starting at 54 MHz is a band for broadcast television channels 2 - 6, then a band for analog FM radio, followed by a band for broadcast television channels 7 – 13. These were commonly referred to as Very High Frequency (VHF) television channels.

The Maritime VHF radio band is at 457 MHz, with narrow channels for analog voice communication.

Television channels 14 – 69, in the bands between 470 MHz and 806 MHz, were called Ultra High Frequency (UHF) television channels.

Attaching a loop antenna to the back of the television set and facing it towards the transmitter, it was possible to detect a radio carrier frequency that was being varied to represent, or be an *analog* of, the value of light intensity detected by a camera scanning left to right, line after line, top of picture to bottom of picture, and repeating.

The radio carrier frequency is varied in time, mimicking the output of the scanning camera as it changes in time. The size of the picture and speed at which the scan is repeated determines the channel bandwidth required.

The energy of the radio waves that are absorbed by the metal of the antenna induces electrons to move in the metal, causing a continuously-varying electrical voltage on wires attached to the antenna. This voltage is again an analog of the light intensity detected by the scanning camera.

The antenna's wires are connected to a *tuner* in the television set. The tuner converts the voltage varying at the radio channel frequencies to voltage varying at normal or *baseband* frequencies.

This voltage is used to vary the intensity of a cathode ray gun: a device that produces a focused beam of electrons pointed at the back of a phosphorous-coated glass screen. The beam is swept left to right across the screen, line by line from top to bottom, then repeat.

Bombarded by electrons, the phosphorous emits light, re-creating on the screen the camera image that is being scanned.

One could imagine families eating TV dinners on TV trays watching Wide World of Disney on Sunday night on a black and white TV in 1966.

FIGURE 3 ANALOG TELEVISION AND PHOSPHOR-COATED SCREENS

2.3.2 Repurposing of Broadcast Television Spectrum

As part of the transition from analog broadcast television to digital broadcast television after the beginning of this century, the frequencies for UHF channels 38 – 69 were repurposed for mobile digital communications.

Because modems are employed, broadcast digital television channels can be packed more densely than channels carrying analog signals without adjacent channel interference. This enabled the movement of existing broadcasters on channels 38 - 69 to the frequencies below channel 37, below 614 MHz, in a process called repackaging.

The band 614 – 698 MHz, formerly UHF channels 38-51, is now referred to as the "600 MHz band". The band 698 – 806 MHz, formerly UHF channels 52-69, is the "700 MHz band".

2.3.3 Two-Way Radio: FDD or TDD

Television is one-way communications. Phone calls and Internet access of course require two-way simultaneous communications, sometimes called *full duplex*.

Using different frequencies for the base station's transmitter and the mobile's transmitter to allow two-way simultaneous communications is called Frequency-Division Duplexing (FDD).

The alternative is to assign the same frequency to the base station and mobile and have them alternate transmitting in time, called Time-Division Duplexing (TDD), also known as *half-duplex*.

2.3.4 600 MHz Band

The 600 MHz band was re-allocated as shown in Figure 4 to create seven paired 5 MHz blocks: 5 MHz for the base station for downloading, and 5 MHz for the handsets for uploading, abbreviated as 5+5.

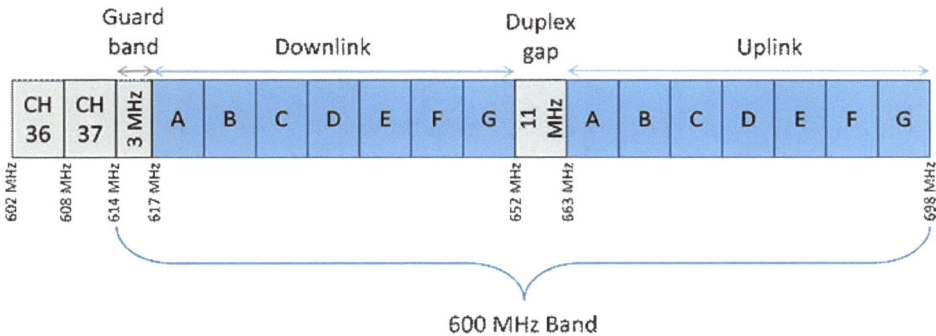

FIGURE 4 600 MHZ BAND ALLOCATION

The number of bits that can be communicated, called the *capacity*, is proportional to the width of the frequency band available for use.

5 MHz of bandwidth is not sufficient for providing broadband wireless, for example, 50 Mb/s per user, to multiple members of the public at the same time. Services at those bit rates are implemented at higher frequencies with channels 20 to 100 MHz wide.

However, radio in the 600 MHz band enjoys transmission characteristics like range and in-building penetration that are far superior to systems at higher frequencies.

This makes the 600 MHz spectrum ideal for Internet of Things (IoT) applications like trackers and monitors, where the data rate requirement is low and wide coverage and excellent in-building penetration is desired.

2.3.5 700 MHz Band

After being repurposed from UHF channels 52 – 69, the 700 MHz band was organized into the Lower 700 MHz band and the Upper 700 MHz band, due to differing statutory requirements, and differing levels of incumbency, as there were approximately ten times more existing broadcast license holders in the lower than the upper.

In North America, the Lower 700 MHz band was reallocated to three blocks of 6+6 MHz plus two unpaired blocks. Fixed, mobile and broadcast services are permitted in this spectrum.

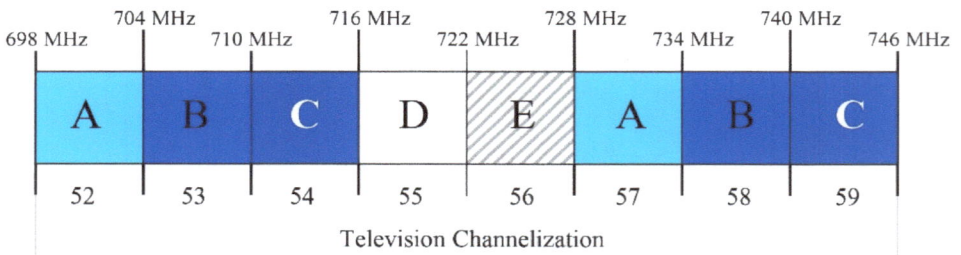

FIGURE 5 LOWER 700 MHZ BAND

In the rest of the world, (called "Europe" in the telecom business), the increasingly-inaccurately-named Third Generation Partnership Project (3GPP) defines different bands: band class 12, which covers the A, B and C blocks; and band class 17 which covers B and C only.

Allowing all carriers' customers to roam on a 700 MHz spectrum allotment was a condition for licenses in both the US and Canada. The Canadian regulator specified this in a document sensibly titled "Conditions of License for Mandatory Roaming and Antenna Tower and Site Sharing and to Prohibit Exclusive Site Arrangements".

The Upper 700 MHz band was allocated as matched blocks of upload and download spectrum, with Block C at 11+11 MHz, Block D 5+5, Blocks A and B at 1+1, plus 12+12 MHz for Public Safety.

757		763			775			787		793			805
			Public Safety								Public Safety		
C	A	D	Broadband	G B	Narrowband	B	C	A	D	Broadband	G B	Narrowband	B
CH. 60	CH. 61	CH. 62	CH. 63	CH. 64	CH. 65	CH. 66	CH. 67	CH. 68	CH. 69				
746	752	758	764	770	776	782	788	794	800	806			

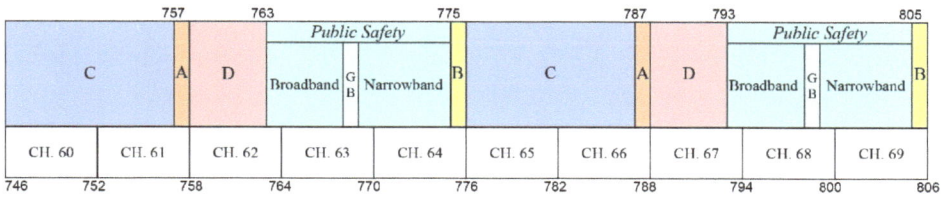

FIGURE 6 UPPER 700 MHZ BAND

During the 9/11/2001 terrorist attack on the United States, it became clear that there were significant problems with mobile communications for public safety personnel: it was very difficult to get cellular service during the emergency; it was difficult if not impossible to communicate between different agencies like FBI, fire and police; and they could not roam on each other's private radio systems.

Telecommunication networks are *overbooked*: there are far more users than there is capacity to serve them all at the same time. This is called *statistical multiplexing*, exploiting the fact that not all users will be communicating at any given time. Users get the possibility of communicating, for a lower price than were capacity provisioned for each user whether they were transmitting or not.

The amount of overbooking is based on historical demand statistics. However, during a crisis, when everyone does try to use their wireless service at the same time, the demand is far higher than usual and there is a capacity failure.

Public mobile communications networks can not be relied upon during an emergency.

The solution was to build a nationwide separate network for public safety use, where it was possible to control how many people have access and thus avoid loading failure. The US Congress required the reservation of the 12+12 Public Safety block in the Upper 700 MHz band nationwide, and created an organization called FirstNet to manage the construction and maintenance of a nationwide cellular network, paid for by taxpayers, only for use by first responders and government agencies.

2.3.6 800, 900, 1800 and 1900 MHz bands

The 800-MHz band was initially allocated for the first generation of mobile cellular in North America, ending up with two 10+10 blocks plus two 5+5 blocks within 824-894 MHz.

In Europe, the 900-MHz and 1800-MHz bands were allocated for second generation mobile called GSM: The Global System for Mobile Communications.

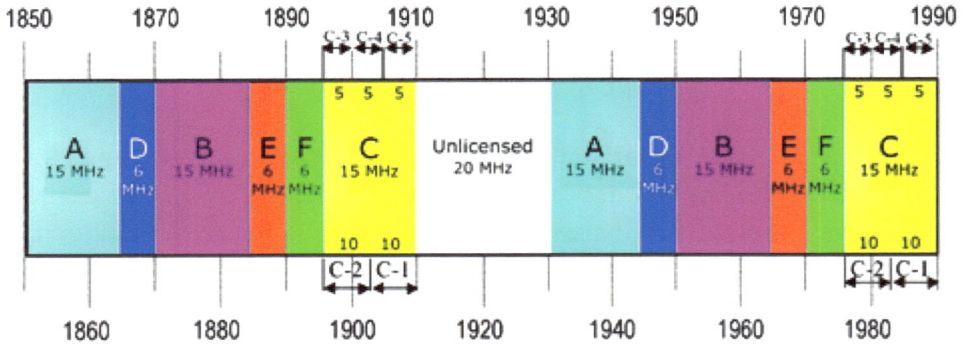

FIGURE 7 1.9 GHZ "PCS BAND"

In North America, the 1900 MHz, or 1.9 GHz band, which actually encompasses 1850 MHz to 1990 MHz, was initially allocated for the second generation of cellular, called Personal Communications Services as depicted in Figure 7, with 15+15, 10+10 and 5+5 blocks.

New bands are allocated, and older bands are re-allocated to new generations of systems on an ongoing basis.

These bands are now allocated for 3rd-generation CDMA and 4th-generation LTE. In the future, 5G will run on all bands.

2.3.7 Unlicensed Bands

The 900-MHz band is an *unlicensed band* in North America, which essentially means it is allocated to the public, and a license to emit energy at that frequency is not required. These bands are also called Industrial, Scientific, and Medical (ISM) bands. The 900-MHz band was used for cordless phones and baby monitors in North America.

At 2.4 GHz is another ISM band. This band is used by numerous technologies including Bluetooth, Wi-Fi, cordless phones, private point-to-point transmissions and... microwave ovens, to avoid the need for a license. This makes the 2.4 GHz band noisy.

| FRIGIDAIRE | ELECTROLUX HOME PRODUCTS | (UL) LISTED |
| HOUSEHOLD MICROWAVE OVEN | INC CHARLOTTE, NC 28262 | 54NJ |

| MODEL NO. FFMV1645TB | SERIAL NO. KG82001262 | MANUFACTURED May, 2018 |
| Vac/Hz: 120/60 Rated Input: 1550W ↓ Rated Output: 1000W Microwave Frequency: 2450MHz | For Customer Assistance Call Support at 1-800-374-4432 www.frigidaire.com | FCC ID : VG8EM044KYY MADE IN CHINA |

THIS PRODUCT COMPLIES WITH DHHS RULES 21 CFR SUBCHAPTER J. Not for built-in installation.
PN:16070000B08767

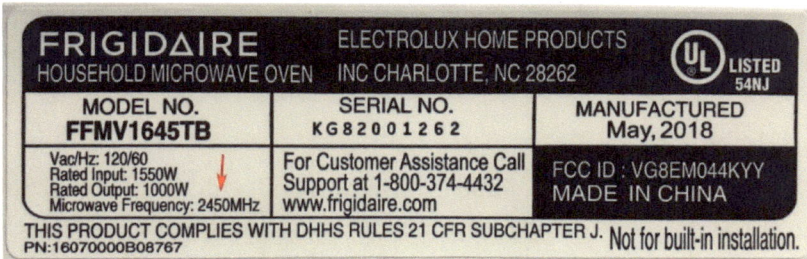

FIGURE 8 MICROWAVE OVEN OPERATING IN THE 2.4 GHZ ISM BAND

Bands starting at 5 GHz are part of the Unlicensed National Information Infrastructure (U-NII). U-NII bands between 5 and 6 GHz are used by 802.11a Wireless LANs, and between 6 and 7 GHz are used by 802.11ax.

Energy at 5 and 6 GHz encounters significant impairment with transmission through walls and other obstructions. Unobstructed line-of-sight between the transmitter and receiver is necessary to achieve the peak bit rates specified in the standards and marketing materials.

2.3.8 2.5 GHz Band

The 2.5 GHz band has a promising future in broadband wireless, with two 50 MHz-wide channels and one 17.5 MHz channel available for mobile and fixed applications.

The channel width has increased from 10 and 15 MHz in the 800 and 1900 MHz bands to 50 MHz, dramatically increasing the capacity available.

2.3.9 3.5 GHz Band

New spectrum from 3550 to 3700 MHz is beginning to be used in North America for Fixed Broadband Wireless services, providing broadband wireless Internet to residences and businesses as an alternative to cable, DSL, fiber or a cellular data plan.

This service is provided from cell towers, and currently uses LTE, but it is not accessible by mobile phones.

The 150 MHz of spectrum is divided into 10 MHz blocks. In the USA, licensees can aggregate up to four blocks to create 40 MHz channels.

However, as frequency increases, transmission characteristics like penetration through obstacles worsens. 3.5 GHz energy is significantly blocked by near-field interference, usually trees in front of the antenna.

Line of sight between the tower and the antenna at the customer premise is required to achieve broadband bit rates.

2.3.10 3.7 GHz C-Band 5G

The 3.7 GHz band (3700 – 4200 MHz) was allocated to Fixed Satellite Service (FSS) space-to-Earth communications, and Fixed Services (FS) using 20+20 MHz for point-to-point long-haul analog radio telephone trunk carrier systems.

Paired with the 5.925-6.425 GHz band (Earth-to-space), these frequencies are known as the Conventional or C-band in the satellite business.

By repacking the incumbent license holders in the 3.7-4.2 GHz band into the upper 200 MHz of the band, 280 MHz of spectrum between 3.7 and 4 GHz has been cleared and reallocated to mobile and fixed communications for broadband terrestrial 5G.

It is allocated in 20 MHz blocks, which can be aggregated up to 100 MHz channels.

Interference with radar altimeter systems was a significant cause for concern in the weeks before the rollout of 5G in the C-band.

Not only would the 5G licensees and the altimeters interfere, it is now illegal for aircraft to emit radio at those frequencies in the USA.

A number of foreign carriers, who claimed they didn't hear about the reallocation, were still using the now-cleared 3.7 – 4 GHz band, and urgently requested 5G rollout at airports be stopped. It wasn't.

Such aircraft cannot use their radar altimeters in the USA until they are brought into conformance with FCC regulations, preventing landing in zero-visibility conditions. The nonconforming altimeters were replaced.

2.3.11 Millimeter-Wave Bands

The FCC has completed spectrum auctions in the 24 GHz band, the 28 GHz band, and the upper 37 GHz, 39 GHz, and 47 GHz bands, representing almost 5,000 MHz (5 GHz) of spectrum for 5G… 200 times more than the bandwidth that was allocated for 1G.

The bands around 30 GHz are called *millimeter-wave* bands, since the wavelength of the energy is measured in millimeters. At these frequencies, molecules in the air block energy, a phenomenon called *atmospheric absorption*, which limits the useful range to perhaps 150 meters.

One application for 5G is data communications for assisted-driving cars, where a central controller assembles cars on the highway into convoys or *platoons* to increase density and reduce energy usage.

Since line-of-sight is required and atmospheric absorption limits range to hundreds of meters, full coverage on a highway at 30 GHz would require radio base stations installed on every second or third streetlight all the way along a highway.

3 Analog Radio

With this lesson, we begin to look at how information can be represented using radio, starting with analog techniques.

3.1 Definition of Analog

The AM radio in an automobile is an *analog* system. We usually define "analog" as "direct representation."

On telephone lines, a voltage is varied up and down as a direct representation of sound pressure waves coming out of the speaker's throat. This voltage must vary within the voiceband frequency constraints at 300 - 3300 times per second, or Hertz (Hz).

3.2 Carrier Frequency for Radio

"Radio," means systems where the energy is vibrating at frequencies measured in the hundreds of millions and billions of Hertz. Therefore, the voltage cannot be varied up and down as a direct representation.

Instead, it is possible to pick a single pure frequency, a wave oscillating at a fixed rate, called a *carrier frequency* and vary or *modulate* its strength up and down as a direct representation of the sound pressure waves coming out of the speaker's throat as illustrated in Figure 9.

3.3 AM, FM and PM

Since amplitude is the technical term for signal strength, the technique of varying the carrier signal strength as a direct representation of the source or *baseband* signal is called Amplitude Modulation (AM). The *envelope* of the varying carrier is the same as the voltage analog.

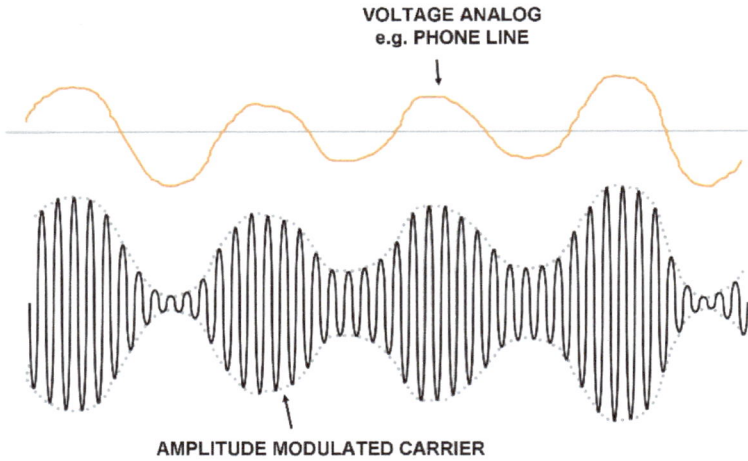

VOLTAGE ANALOG
e.g. PHONE LINE

AMPLITUDE MODULATED CARRIER

FIGURE 9 THE ENVELOPE OF AN AMPLITUDE-MODULATED CARRIER IS
AN ANALOG OF A BASEBAND ELECTRICAL SIGNAL

Other radios use continuous modulation of the frequency of the carrier,
called Frequency Modulation (FM), as it is less sensitive to noise.

It is also possible to continuously modulate the *phase* of the carrier, that is,
varying the relative position in time of the carrier as a direct representation
of sound strength. This is called Phase Modulation (PM).

4 Digital Radio: Modems

When the information to be communicated is not a continuously-varying phenomenon, like the strength of the sound coming out of someone's throat, but rather 1s and 0s, a different tactic is adopted to vary the carrier: shifting, or *keying* between discrete states.

4.1 Amplitude Shift Keying

The simplest example is *Amplitude Shift Keying (ASK)* where a single carrier frequency is used, but instead of varying it continuously, two amplitudes are defined, and the output signal is shifted, or *keyed* back and forth between the two amplitudes to represent the 1s and 0s.

One volume or amplitude is used to represent a "1," and another amplitude is used to represent a "0." Sometimes, one of the amplitudes is zero.

| Binary Code | 0 1 0 1 1 0 0 0 1 0 |

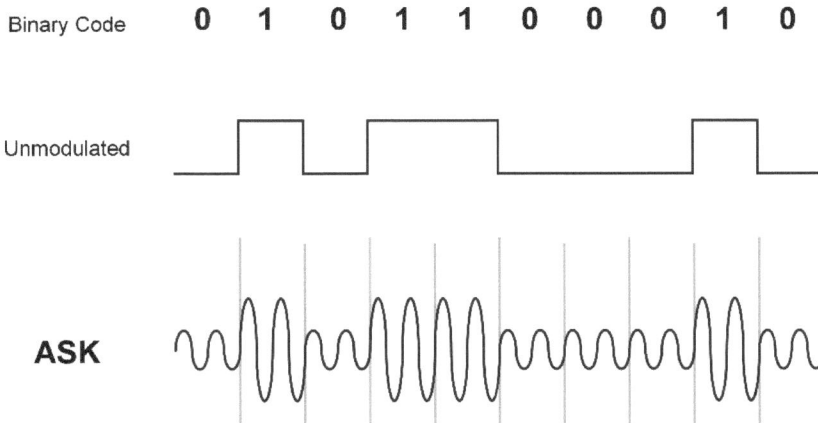

FIGURE 10 "KEYING": SHIFTING AMPLITUDE BETWEEN TWO CHOICES

Note that this is not continuous *amplitude modulation* (AM) as is used on radio stations. At the radio station, a carrier frequency of something like 800 kHz is selected, and the amplitude at that frequency is varied continuously to represent the sound pressure waves coming out of the disk jockey's throat.

In contrast, ASK uses only two amplitudes. One amplitude means "0," and the other one means "1."

This technique is susceptible to noise. Most noise, like that from microwave ovens and fluorescent lights, is *additive*. Noise adds to the signal and pushes the amplitude over the boundary line between what means 0 and what means 1, and so causes an error.

Thus, this technique is not very good for physical media which have added noise like copper wires and radio. It is useful, however, for physical media which don't have added noise, like fiber optics. This is how 1s and 0s are represented on fibers on many systems. Light on = "1" and light off = "0" on Optical Ethernet.

4.2 Frequency Shift Keying

An improvement on Amplitude Shift Keying is Frequency Shift Keying (FSK), where the amplitude is constant, and two carrier frequencies are used. To indicate a "0," one of the frequencies is transmitted. To indicate a "1," the other frequency is transmitted.

This reduces the sensitivity of the system to noise, but introduces a different problem: shifting back and forth between two frequencies at a third frequency has the effect of creating energy at the multiples of the frequencies, called *harmonics*, which limits the maximum signaling rate.

4.3 Phase Shift Keying

On most radio systems today, shifting or keying between different *phases* of the carrier is employed. That is, shifting the position of the sine wave illustrated in Figure 11 back and forth in time with respect to a reference. This is more efficient and less susceptible to noise, and is called Phase Shift Keying (PSK).

Binary Code	0	1	0	1	1	0	0	0	1	0

Unmodulated

ASK

FSK

PSK

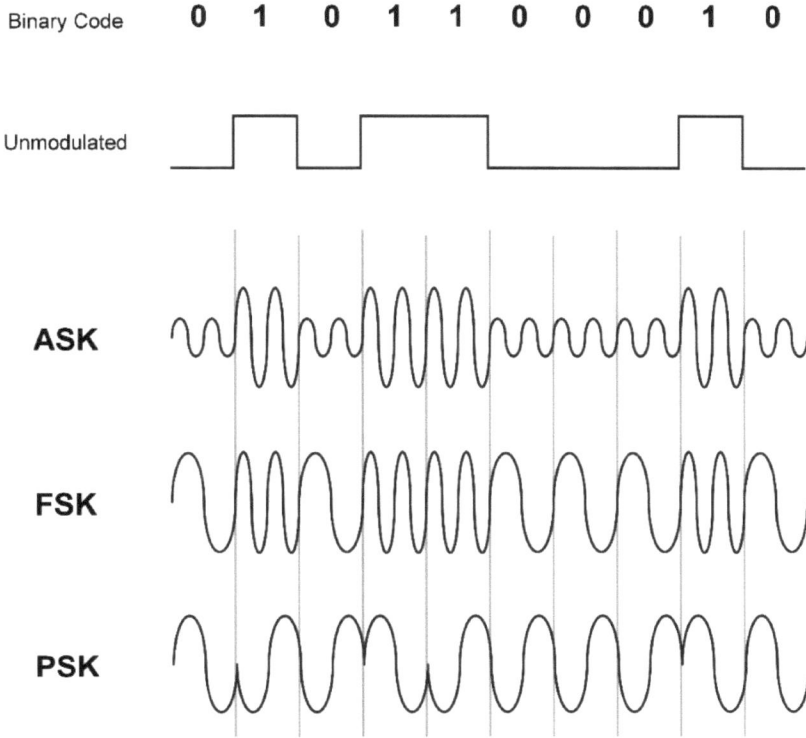

FIGURE 11 AMPLITUDE, FREQUENCY AND PHASE SHIFTING

4.4 Baud Rate

Each state of the modulation is called a *signal*. The number of times per second that the state of the modulation can be shifted to a different value is the number of signals per second.

The number of signals per second is called the *baud rate*, named after the Frenchman Emile Baudot, who invented the precursor to ASCII and channelized Time-Division Multiplexing in the 1800s.

The bit rate achieved equals the baud rate (the number of signals per second) times the number of bits that each signal represents.

In the examples so far, there were two choices: two amplitudes, two frequencies, or two phases. The two choices are called *signals* or *symbols*. One of them is designated to represent "0" and the other represents "1".

Therefore, each time a signal is transmitted, it is representing either a 0 or a 1; one bit is communicated per signal.

To communicate more bits per second, the baud rate can be increased – but only up to a maximum determined by the width of the frequency band available.

The bit rate can also be increased by increasing the number of bits that each signal represents.

4.5 More Signals = More Bits

To indicate more bits per second, more signals are used.

The number of bits indicated by choosing a particular signal is related to the number of signals available by powers of two: $n = 2^b$. This can be demonstrated by assigning a number to each signal and writing the numbers in binary.

For example, four choices for frequencies could be defined for a frequency-shifting system. The signals could be numbered 0, 1, 2 and 3, or in binary as 00, 01, 10 and 11. Then, choosing one of the four frequencies, and transmitting that would indicate 2 bits.

If eight choices for frequency were available, since $8 = 2^3$, choosing one out of eight and transmitting that frequency specifically would indicate 3 bits. If 64 different frequencies were defined, $64 = 2^6$, so choosing one frequency out of the 64 available choices and transmitting it would indicate six bits in one fell swoop.

4.6 QPSK: 2 Bits per Signal

Illustrated in Figure 12, a modulation scheme with four choices for phases is called Quadrature Phase Shift Keying (QPSK).

Each of the four phase shifts can be given a number: 0, 1, 2, and 3 – or 00, 01, 10 and 11 in binary. Then each time a signal is sent, that is, each time the transmitter shifts the phase of the transmitted carrier to a new value, that represents two bits.

The receiver, measuring the phase of the received carrier, determines which of the four phase shifts is happening, and outputs the two-bit number corresponding to that phase shift.

Binary Code **0 1 0 1 1 0 0 0 1 0**

Unmodulated

QPSK

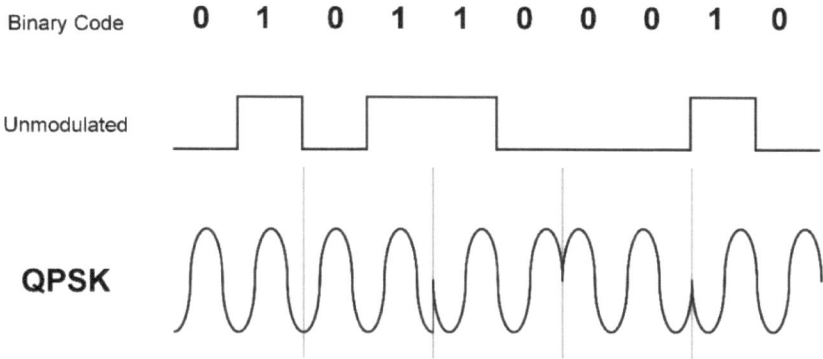

FIGURE 12 QPSK: EACH SIGNAL REPRESENTS TWO BITS

While keeping the baud rate the same, the four-phase QPSK scheme communicates twice as many bits per second as the two-phase PSK.

4.7 QAM

To define even more signals, a scheme combining amplitude shifting and phase shifting called Quadrature Amplitude Modulation (QAM) is used.

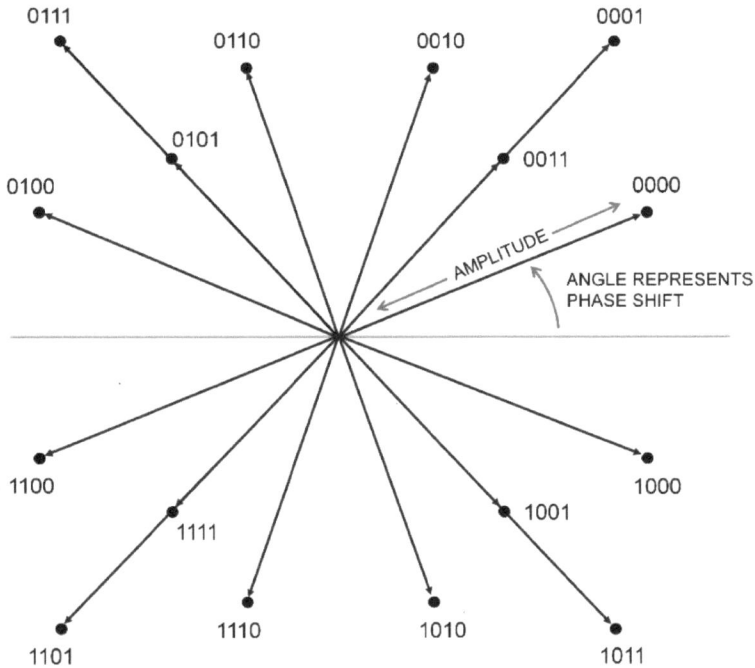

0111   0110   0010   0001

0101   0011

0100   0000

AMPLITUDE

ANGLE REPRESENTS
PHASE SHIFT

1100   1000

1111   1001

1110   1010

1101   1011

FIGURE 13 QAM-16 PHASOR DIAGRAM

On paper, arrowed lines with a particular length and a particular rotation called *phasors* are used to visually represent the QAM signals in *polar coordinates* rather than the familiar x,y *cartesian* coordinates.

The length of the arrow represents the signal's amplitude, and the rotation angle represents its phase shift from the previous signal.

Figure 13 illustrates QAM-16, where 16 different combinations of phase and amplitude shifts are defined. As can be seen in the diagram, numbering each of the 16 signals in binary requires 4 bits, since $16 = 2^4$.

Therefore, each signal represents a 4-bit number, and the bit rate is four times the baud rate.

QAM-64 is eight phase shifts and eight amplitude shifts, defining 64 signals, each of which represents six bits, since $64 = 2^6$.

4.8 Limits

The number of signals can not be increased indefinitely, since as more signals are defined, the differences between them become smaller, and the receiver's decision-making more error-prone.

Noise, picked up along with the received carrier signal, causes errors deciding what amplitude and phase of carrier is being heard at the receiver. When a decision error is made, no bits are communicated.

4.9 Summary

Bits are represented by transmitting energy at a single frequency, called a *carrier frequency*, and in QAM using specific amplitudes and phases of the carrier to represent numbers.

The transmitter and receiver agree in advance the allowed combinations of phase and amplitude, and what number each combination represents.

The receiver measures the amplitude and phase of the carrier, and yields the number that combination represents.

The number of bits per second is equal to the number of signals per second times the number of bits represented by each signal.

A *signal* in QAM is transmitting a carrier with a specific amplitude and phase for a specific length of time.

The number of signals per second is called the *baud rate*.

In QAM, the baud rate is how often the phase and amplitude can be changed to a different combination.

This is limited by the width of the frequency band available, defined by the width of the *channel* in which the carrier is being transmitted.

The second part of the equation, the number of bits represented per signal, is determined by the number of different signals that the transmitter might send.

In QAM, the number of signals is the number of specific combinations of phase and amplitude defined.

The number of signals that are practical to use in a given application is determined by the received carrier signal strength, and by noise.

The more signals defined, the smaller the differences between them.

Conversely, as noise increases, the variation of the received signal becomes bigger.

If the variation is more than the difference between adjacent signals, the receiver makes a decision error and no bits are communicated.

5 Propagation, Penetration and Fading

5.1 Propagation

No-one is sure how radio waves travel through space, nor indeed why they choose to do so; we can only measure and characterize how and when they do this.

In regard to radio waves, *propagation* is the term that is used instead of "travel." The radio wave is electro-magnetic energy vibrating at, or near the carrier frequency, for example, 2.4 billion times per second.

In a radio system, first, electricity vibrating at this frequency is produced on copper wires. This electricity is conducted through an antenna, causing electro-magnetic energy to emanate from the antenna and propagate away at the speed of light.

5.2 Omni Antennas

An "omni," or omnidirectional, antenna radiates energy equally in all directions at right angles to the antenna. An example is the pencil-sized antennas on Wi-Fi access points, which radiate energy in a donut shape outward from the sides of the antenna.

To provide coverage on the same floor of the building, the antenna would be positioned vertically so that it radiated energy horizontally. Likewise, to provide coverage on upper and lower floors, the antenna might be tilted at a 45 or 90-degree angle to radiate some or all the energy in the vertical direction.

Another example of an omni antenna is the stick-shaped antennas on a tower depicted in Figure 14.

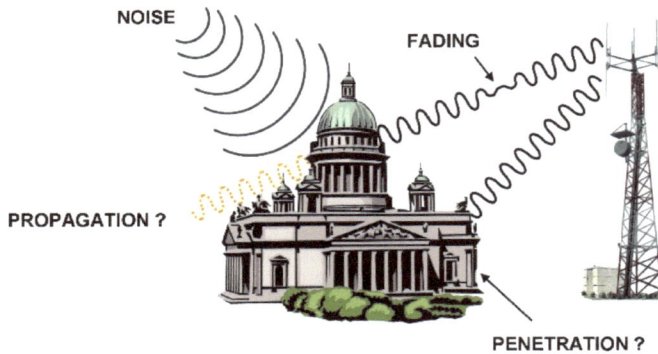

FIGURE 14 PROPAGATION, PENETRATION AND FADING

5.3 Directional Antennas and Sectorization

Figure 14 shows omni antennas mounted on a tower as part of a cellular radio system. To increase *capacity*, that is, more users and/or faster bit rates, *directional antennas* are used.

Directional antennas are constructed so that they radiate most of their energy in an arc, only covering part of the 360-degree circle around it. This allows multiple directional antennas to be installed, each facing a different direction, and each providing an independent radio signal within its arc. This is called *sectorization*.

5.4 Attenuation and Carrier-to-Noise Ratio

To understand the coverage of a radio system, i.e., where it can be successfully used, we are interested in understanding how far the waves travel, and how well they penetrate into buildings.

In the absence of obstacles, the waves travel forever, to the end of the Universe and beyond, at the speed of light.

In practice, to determine if a particular location is covered by the radio system, it is the strength of signal that can be detected versus the amount of undesired energy added to the signal, called noise or interference, that is of interest. This is called the *signal to noise ratio* (SNR).

To eliminate confusion between the total measured power at the carrier frequency versus the tiny fraction of that which is the power of the signal modulating the carrier, the term Carrier to Noise Ratio (C/N), or CNR is often used instead of signal-to-noise ratio.

The power that a receiving antenna detects is *attenuated*, or lessened, in proportion to the square of the distance away from the transmitter due to the ever-expanding area of the sphere the transmitted power covers, while the receiving antenna stays the same size.

Obstacles absorb some or all the energy, turning it into heat instead of allowing it to continue propagating.

The amount of power lost due to an obstacle is dependent on the frequency of the carrier; the composition of the material: plaster, concrete, brick, wood, metal; on how much energy has been able to refract (bend) around the object; and how much energy has reflected off other obstacles to arrive at the measurement point via an indirect path.

Transmission from the outside of a building through walls, called *in-building penetration*, is of particular interest for public mobile cellular systems.

Lower frequencies penetrate through walls and other obstacles better than high frequencies. This is the reason why fog horns are very low frequency.

5.5 Fading

The carrier can take many paths to arrive at the receiver, reflecting off nearby objects like buildings and cars. Each of these paths is a different length, so copies of the signal arrive at different times, shifted in *phase*.

A reflection arriving half a wavelength later than the main carrier is called *destructive interference*, and cancels out the main carrier resulting in a detected power of zero. In practice, there are many reflections arriving at different times, so the detected power varies. This is called *fading*.

When the environment has moving obstacles (cars and tree leaves for example), the fading changes rapidly. Measures like Forward Error Correction, where extra error-correcting bits are transmitted so the receiver (or forward end) can correct errors, are required to compensate for signal fading to achieve high transmission rates.

5.6 Interference

Interference from other radio sources (including fluorescent light transformers, other transmitters, and the sun) is the other side of the C/N ratio.

In designing a radio system, the actual noise power at the frequencies of interest is of critical importance. Engineers use the term *spectrum clearing* to refer to the relative amount of noise at specific frequencies in specific areas, and sometimes the process of identifying and removing sources of noise.

Course 2232

Mobile Communications

1 Mobile Network Components and Operation

5.7 Mobile Network and Mobility

"Mobile network" is the term given to distributed radio systems, designed so that many users who may or may not be moving around can share a radio band and communicate amongst themselves, and to other people or computers on wired networks like the telephone network and the Internet.

Communications takes place between the user's mobile radio and a fixed radio base station, that is in turn connected to other base stations and to the wired networks.

Mobility means that if a mobile moves too far away from its base station, causing signal loss, communication is not lost, but maintained by having the mobile communicate with the next base station further down the road.

5.8 Handset, SIM Card and IMSI

The cellphone, also called a *mobile*, terminal, smartphone or handset contains the components necessary for communication including a microphone, speaker, codec, screen, radio and battery.

The mobile has a hard-coded International Mobile Equipment Identity (IMEI) to identify the hardware. The mobile also contains a removable Subscriber Identity Module (SIM) card, that has a unique number called the International Mobile Subscriber Identity (IMSI), plus one half of an encryption key pair used to authenticate the SIM data.

During operation, the mobile learns the identity of the particular base station it is communicating with, and stores this information as the Location Area Identity on the mobile's SIM card.

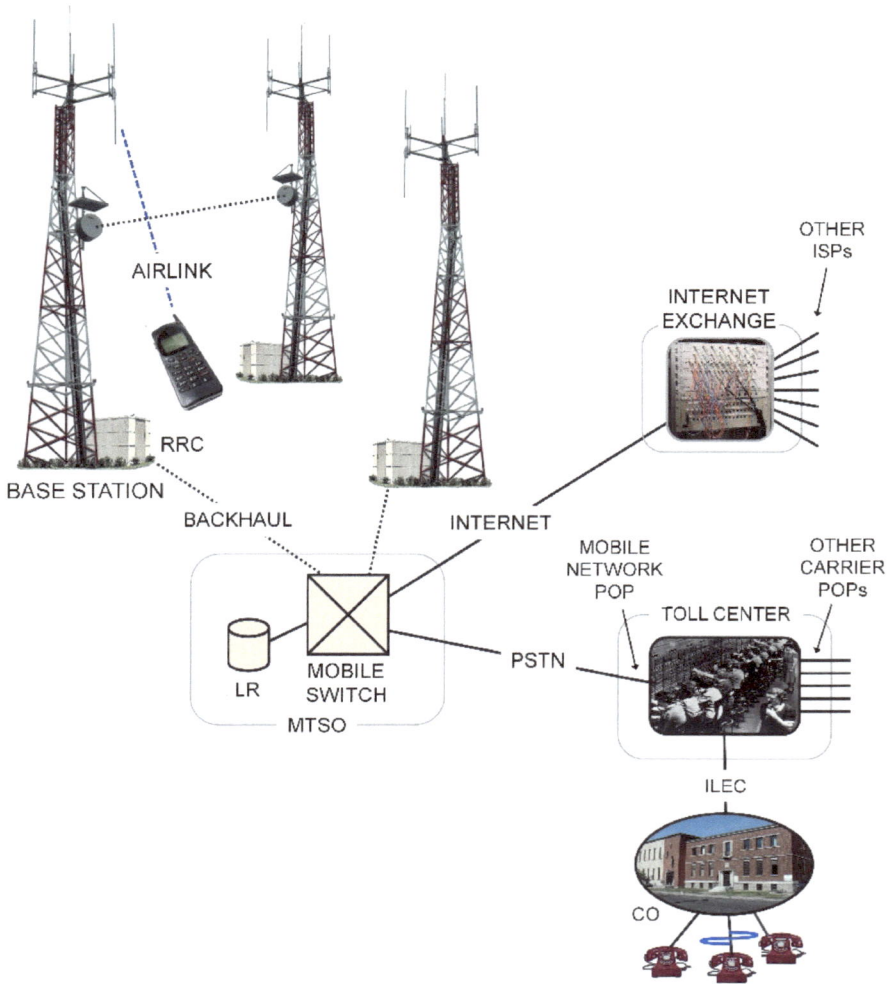

FIGURE 15 COMPONENTS OF A MOBILE NETWORK

5.9 Airlink, Base Station, Towers and Cells

As illustrated in Figure 15, the mobile is connected to the network via an *airlink* to a *base station*, which is called an evolved NodeB (eNB) beginning with the fourth generation.

The base station includes the Radio Resource Controller (RRC) that schedules communications between devices and the network, allocates capacity, controls the signal power used to communicate, negotiates the power state of each device and more.

The physical base station enclosure also contains racks of radio transmitter-receivers or *radio transceivers*, which produce and detect energy at radio frequency in the form of electricity, and convert between radio frequencies and normal or *baseband* frequencies.

The transmitters are connected with thick coaxial cables to *antennas* that convert the electricity at radio frequency to electromagnetic waves at the same frequency, which then propagate away at the speed of light.

The antenna also captures energy of electromagnetic waves impinging upon it, inducing electricity to flow on the antenna, that is fed to the receivers.

The mobile is similarly equipped with an antenna and radio transceiver.

Usually, different frequencies are used for transmitting and receiving to allow simultaneous two-way communications without interference, known as Frequency-Division Duplexing (FDD).

A physical support for the antennas such as a *tower* or building is another element of a base station.

The area on the ground covered by the base station is a *cell*.

5.10 Mobile Telephone Switching Office

The base stations are connected to a *mobile switch*, in a building called a Mobile Telephone Switching Office (MTSO).

A mobile switch is a telephone switch, an IP-based softswitch, with additional capability to keep track of users moving between base stations. *Location Register* (LR) is the general term for a database that keeps track of user status and location via the International Mobile Subscriber Identity and the Location Area Identity from the SIM card.

On power-up, the handset *registers* with the switch, which records the ID of the base station the handset is using in the LR. When the handset is moved, the value of the Location Area Identity will change and the handset initiates a location update to the LR.

5.11 Backhaul and Network Connections

In many cases, the base station will be connected to the mobile switch with fiber. Otherwise, a point-to-point microwave link will be used to connect one base station to another that has a fiber link to the mobile switch.

This part of the network is called the *backhaul*.

Delivering broadband wireless to many people requires fiber backhaul.

As illustrated in Figure 15, the mobile switch is connected to the Public Switched Telephone Network (PSTN), to allow calls between cell phones and landlines. This happens at a building traditionally called a Toll Center or Class 4 switching office.

The mobile switch is also connected to the Internet, that is, to other Internet Service Providers (ISPs) at a building called an Internet Exchange (IX).

5.12 Incoming Call and Paging

When there is an incoming call, the mobile switch will *page* the handset from the base station(s) corresponding to the Location Area Identity value, a short message to find out if the handset is still operating and near that base station.

If the handset does not answer the page, the network will resend the page on neighboring base stations in the area, or, in some cases, to all of the base stations on the network.

Once the handset answers the page, the base station will transmit the Caller ID and tell the handset to start "ringing".

When the user presses the "talk" button, voice communications take place from the mobile to the base station over the airlink, then from the base station to the mobile switch via the backhaul.

For a mobile-to-mobile call, the communications will be routed to a base station. For mobile to landline, the call will be routed to the PSTN.

5.13 Mobility and Handoffs

If a user moves during a call, at some point, the user will be *handed off* from one base station to another.

This means that the network will switch to using a different base station to communicate to the handset, and may involve changing the radio frequencies used by the mobile and base station.

The handoff implements mobility: the ability to maintain communications while moving.

6 Cellular Principles

6.1 Coverage, Capacity and Mobility Requirements

The first mobile radio system connected to the PSTN in North America was called *MPS*: The Mobile Phone System. MPS employed base stations in large metropolitan areas and radios in automobiles with big whip antennas. The caller had to call a "mobile operator" and ask for a specific "mobile number," and would (maybe) be patched through.

The geographical areas where service was available – the *coverage* – were very limited. There was very little *capacity*, so not many people could use the system at the same time, and ironically, it did not support *mobility*.

Once the call was patched through, if the person with the mobile radio drove too far away from the base station, the call would be dropped.

To meet the requirements of coverage, capacity and mobility, *cellular* radio systems were deployed.

6.2 First Generation

The first generation of cellular, called the Advanced Mobile Phone System (AMPS) in North America (the improvement on MPS), employed Frequency-Division Multiplexing where the allotment of radio frequencies would be divided into narrow radio channels.

Since multiple users could access the system, it was referred to as Frequency-Division Multiple Access (FDMA).

A radio frequency band, or spectrum, was allocated for this service by the federal government between 800 and 900 MHz in North America, and above 900 MHz in many other countries.

In North America, an allotment was given to an affiliate of the incumbent telephone company in each market (which, in a wonderful piece of jargon was called the *wireline cellular*) and an allotment was given to a competitor.

6.3 Cellular Design to Meet the Coverage Objective

A Mobile Network Operator (MNO), also known as a cellular carrier, would need a real estate department to find locations where they could construct the base stations. In the first generation, this included large towers to support the antennas.

They would divide their allotment of frequencies into seven smaller blocks of frequencies called groups.

FIGURE 16 CELLULAR RADIO FREQUENCY DESIGN

Then, at a base station, the operator would tune the base station transceiver to use frequencies within one of the seven groups; the radio coverage area around the tower would be something like 3 miles or 5 km in radius.

On the ground, this is the *cell*. Figure 16 illustrates a cell labeled "1" centered on Menlo Park, California in Silicon Valley.

Then, the operator would find another location perhaps six miles or ten kilometers away and build a second base station using a second group of frequencies... for example, in Fremont across San Francisco Bay.

This pattern is continued to build seven base stations, using all seven groups within the allotment. At that point, the operator would have coverage in the geographical area covered by cells 1 to 7 as illustrated, all the way over to Cupertino, California where Apple is headquartered, and all their allotted spectrum would be used.

6.4 Frequency Re-use

The operator could then find another geographic location, for example, near Woodside, California, where Neil Young has a 1500-acre ranch in the middle of some of the nicest real estate on the planet.

They could then convince the locals to let them build an eighth tower, where they could, in this example, re-use frequency group 7.

Since the eighth tower is in Woodside, more than 20 miles away from Cupertino, and the radios have a relatively short range, the same frequency group can be re-used, and the base stations will not interfere with each other.

This is the idea behind a cellular radio system: being able to **re-use** the same frequency groups repeatedly in different geographic locations to meet the coverage requirement.

6.5 Handoffs

To implement *mobility* when a user moves too far away from a base station, they must be handed off to another base station without having their phone call dropped.

7 1G: Analog Frequency-Division Multiple Access

In this lesson, we take a closer look at the technology for the first generation of mobile cellular. Though of course 1G is long obsolete, the basic concept of frequency-division multiplexing returns in 4G and 5G, making understanding 1G a first step in understanding all the rest.

The first radio telephone system was called the Mobile Phone System (MPS) in North America. It required users to install large antennas on their cars, and callers had to call a special mobile operator to be manually patched through to the mobile.

7.1 AMPS, NMT and TACS

The subsequent improvement, the first generation (1G) cellular system deployed in North America, was called the Advanced Mobile Phone System (AMPS). Similar technologies, including NMT and TACS, were deployed elsewhere in the world.

7.2 Frequency-Division Multiplexing

In North America, each service provider ended up with an allocation 25 MHz wide in the 800 MHz band (for example, 824-849 MHz uplink and 869-895 MHz downlink).

The allocation was divided into 30 kHz channels. Groups of 45 channel pairs were assigned to base stations and radios connected to antennas on towers, serving an area about 3 miles or 5 km around the tower: the cell.

7.3 Frequency Re-Use

The groups of channels were assigned to base stations so that the channels could be re-used at other base stations geographically far enough away so they would not interfere with each other.

Organizing the channels into 7 groups was referred to as N=7, and allowed coverage of arbitrarily large areas using a honeycomb pattern for the cells as illustrated in the previous lesson.

7.4 Analog FM

When communications were initiated, a user would be assigned a pair of channels: one for transmit and one for receive.

The radios implemented analog FM, which is a carrier centered in the channel whose frequency is varied continuously back and forth as an analog of the strength of the sound. This is the same technique used on the FM radio in a car.

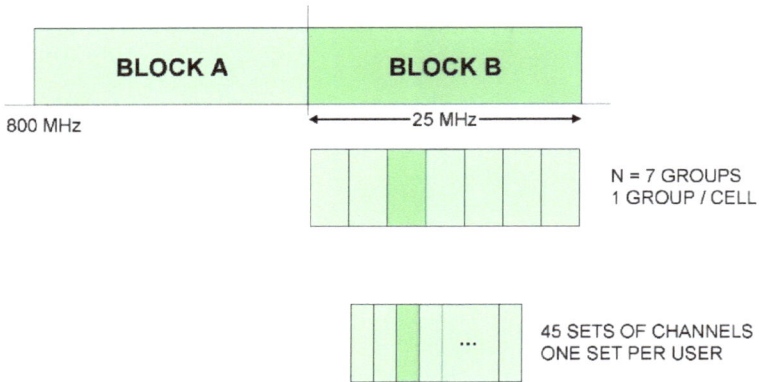

FIGURE 17 FREQUENCY-DIVISION MULTIPLEXING

7.5 Difficulties

7.5.1 Eavesdropping

If the user did not move, and there were not many other users moving about, the user would remain on those channels for the duration of the phone call.

This made it very simple to eavesdrop on conversations because all that is required was an FM radio tuned to the 800 MHz band.

In response, in 1994 the US government required all radios sold in the US to be blocked from receiving the 800 MHz band, though scanners sold prior to this rule, and those sold in other countries had no such restriction.

7.5.2 Modem Disconnect During Handoff

Data communications over this system was similar to data communications using a dial-up modem on a telephone line. In this case, an external modem would be connected to the cell phone, which would establish a phone call to a far-end landline with a modem attached. The two modems would signal 1s and 0s using carrier frequencies.

If the user moved too far from the base station, a handoff to another base station would occur.

Since each base station uses different frequency groups, this meant that two things had to happen. First, the handset would have to change radio channels, and second, both the handset and switch would have to start communicating to the new base station.

If the system worked perfectly, this took 0.2 seconds. During this 200 ms handoff period, communications end-to-end would cease, called *muting* in the cellphone business.

During a voice call, the communications muting or dropping out for 200 MS sounds like a click. However, if modems were communicating, the carrier signal would be lost during the 200 ms handoff and the modems would cease communicating.

This meant that data communications were unwieldy and unreliable. Even stationary, a handoff could happen if someone in handed into a cell, another user might be handed off to a neighboring (overlapping) cell to make room.

7.5.3 Low Capacity

Another problem with AMPS was low capacity. 45 channel pairs per cell meant, in practice, that there were 40 users per cell, since it was necessary to keep some open for users driving into a cell.

40 users in a cell with 3 miles radius around the tower is a paltry 1.5 users per square mile.

Sectorization, or using multiple antennas each with shaped beams to create multiple triangular-shaped cells (called *sectors* to confuse people) around the tower were employed to improve capacity; but not enough for the immense popularity of mobile communications.

Better technology was required.

8 Second Generation: Digital

The second generation of cellular technology (2G) employed lower power, smaller cells, and implemented digital communications, meaning the use of modems to communicate digitized speech and data.

The advantage of implementing digital communications is better sound quality, better signaling and control capability, and its inherent ability to communicate data for mobile access to the Internet.

FIGURE 18 2G CARRIERS

2G was initially called Personal Communication Services (PCS) in North America, and the Global System for Mobile Communications (GSM) throughout the rest of the world.

8.1 Spectrum

While 1G AMPS had been deployed at 800 MHz, second-generation cellular (2G) was initially deployed at 1900 MHz = 1.9 GHz in North America. In the rest of the world, GSM was deployed at 900 MHz and 1.8 GHz.

In North America, handsets were *dual-mode*, meaning they could support both AMPS at 800 MHz and PCS at 1.9 GHz. This was the beginning of a requirement for backward-compatibility.

8.2 Incompatible Spectrum-Sharing Technologies

Several different incompatible technologies were deployed for spectrum-sharing for the second generation.

8.2.1 CDMA: IS-95

In North America, Verizon, Sprint, Bell Mobility, TELUS Mobility and others deployed Qualcomm's CDMA or *Code-Division Multiple Access* technology for 2G following the IS-95 standard.

8.2.2 TDMA: IS-136

Cingular (now AT&T Wireless), Rogers, and others deployed TDMA or *Time-Division Multiple Access* technology following the IS-136 standard. This is sometimes referred to as Digital AMPS or D-AMPS as it was designed to use modems on the same radio channels and same base stations as 1G AMPS.

8.2.3 GSM

In the rest of the world, the GSM standard was deployed. GSM employs TDMA on wider channels than IS-136. GSM became by far the most popular 2G standard worldwide.

The fundamentals of TDMA and CDMA are covered in a following lesson.

9 PSTN Calls Using the Native Phone App: "Voice Minutes"

Putting the discussion of TDMA vs. CDMA aside for a moment, we'll first understand how digital cellular radio works.

9.1 Voice Communication End to End

As illustrated in Figure 19, the cellphone contains a microphone, which creates a voltage that is an analog of the strength of the sound pressure waves at the microphone.

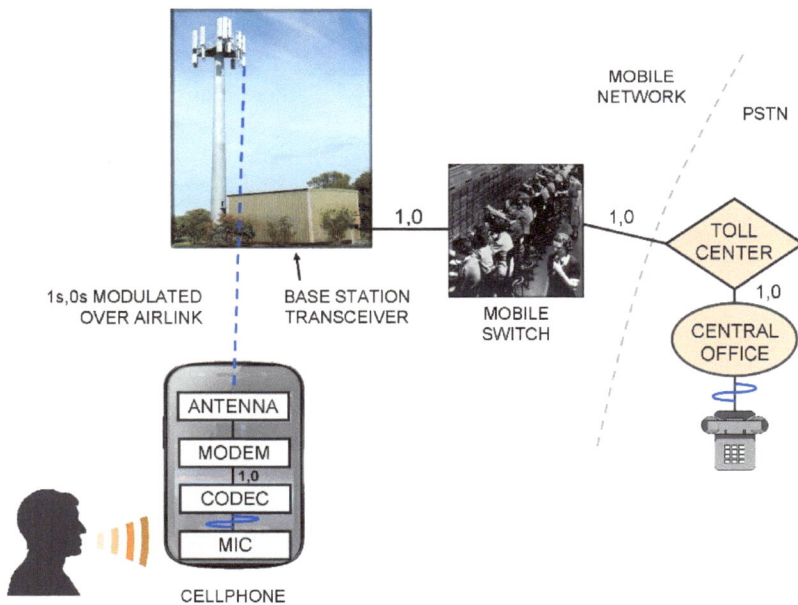

FIGURE 19 VOICE COMMUNICATION OVER DIGITAL CELLULAR

This analog signal is fed into a codec, or *vocoder* inside the phone that digitizes the analog waveform. Additional complex digital signal processing may be performed. The result is 1s and 0s representing the digitized speech.

Then, a modem that operates at radio frequencies is used to represent those 1s and 0s using a modulation technique such as Quadrature Amplitude Modulation within the radio band, in the form of electricity.

This electricity is conducted to an antenna in the phone from which the modem waveform radiates into space in the form of radio waves, or electro-magnetic energy.

At the base station, an antenna converts received radio waves to electricity and feeds them to a modem in the Base Station Transceiver that interprets them to produce 1s and 0s. Complex signal processing is performed on those 1s and 0s to extract the original digitized speech.

This digitized speech is then *backhauled,* or transmitted back to the mobile switch, to be routed to the PSTN for a mobile-to-wireline call, or routed to another base station for a mobile-to-mobile call.

9.2 Coding

The speech is digitized to somewhere between 9 and 13 kb/s for transmission over the airlink, far less than the 64 kb/s DS0 rate used in the traditional landline PSTN.

For mobile-to-wireline calls, the speech must be converted to the format currently required for interconnection between the mobile network and PSTN wireline network, which is DS0 channels, at least for companies that are competitors.

In the past, the coding and formatting standard for interconnect in the PSTN was the G.711 codec at 64 kb/s, carried in a DS0 in DS3s in SONET frames.

In new PSTN systems it is the G.711 codec producing 64 kb/s carried in IP packets in Ethernet frames... however, at present, there are no standards or tariffs for interconnect between competing carriers by exchanging voice in IP packets. The digitized voice in packets is extracted and rendered as a 64 kb/s channel.

In the future, a more efficient coding technique, and/or one that supports a wider frequency band ("HD voice") than the 64 kb/s G.711 codec might become standard.

10 Mobile Internet: "Data Plan"

10.1 "Data" is Internet Traffic

In a quaint, old-fashioned use of terminology, all traffic apart from telephone calls, Short Message Service (SMS) texts, and network messages has been referred to as "data" in cellular billing plans.

Though when this term was first put into use, the screen sizes and bit rates were too low for useful web surfing, things have changed, and "Internet traffic" would be a more accurate term than "data".

Internet traffic of course includes VoIP telephone calls using applications like Skype, email, web pages, social media, videoconferencing, YouTube and Netflix video on demand, Google maps, app downloads and updates, and a million other applications, all communicating between the smartphone and another device over the Internet.

10.2 Using the Built-in Modem

We can use the capability of the digital cellular system that moves digitized speech to move 1s and 0s that are representing anything else. One popular application is connecting a laptop to the Internet.

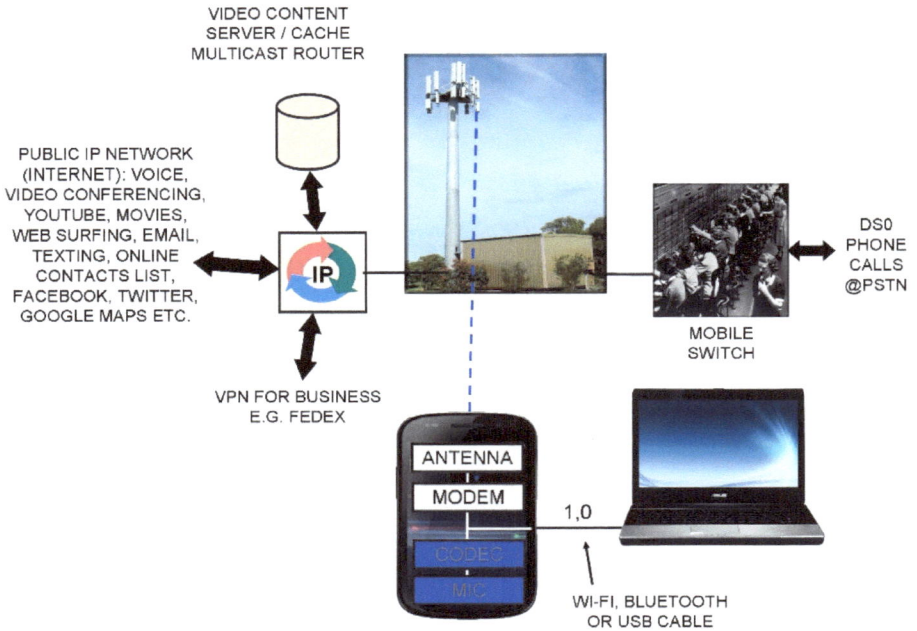

VIDEO CONTENT
SERVER / CACHE
MULTICAST ROUTER

PUBLIC IP NETWORK
(INTERNET): VOICE,
VIDEO CONFERENCING,
YOUTUBE, MOVIES,
WEB SURFING, EMAIL,
TEXTING, ONLINE
CONTACTS LIST,
FACEBOOK, TWITTER,
GOOGLE MAPS ETC.

IP

DS0
PHONE
CALLS
@PSTN

MOBILE
SWITCH

VPN FOR BUSINESS
E.G. FEDEX

ANTENNA
MODEM   1,0

CODEC
MIC

WI-FI, BLUETOOTH
OR USB CABLE

FIGURE 20 MOBILE INTERNET ACCESS

Essentially, the microphone, speaker, screen, and keyboard in the phone are ignored. Instead, the computer connects to the modem in the cellphone and uses that along with the radio, antenna, and battery to communicate IP packets from the computer to the cellular network base station.

From there, the received packets are routed to the mobile switch and relayed to local content servers, the Internet, or to a carrier providing a service with guaranteed quality and security.

A very similar story can be implemented with what the marketing department might call a "stick" – a modem, radio and antenna built into a small dongle that plugs into a USB port on the computer. The "stick" implements the same capability as a cellphone, but without the speaker, microphone, codec, battery, keyboard and screen.

10.3 Tethered Modem

Figure 20 illustrates the modem in the cellphone used to move data from a computer to the Internet via the cellphone. When the cellphone is connected to a laptop with a USB cable, the cellphone is acting as an external or *tethered* modem, i.e., physically tied to the computer.

10.4 Wi-Fi and Bluetooth Links

One could also connect the computer to the phone with a Bluetooth wireless link, which is a 2 Mb/s radio that is separate from the cellular radio, running in the 2.4 GHz ISM band.

The most convenient option once set up, is to activate the feature in a smartphone that turns it into a Wireless LAN (Wi-Fi) access point on the 2.4 GHz and/or 5 GHz unlicensed bands.

The computer then connects to the smartphone's Wi-Fi access point just as it would connect to any other Wi-Fi access point at Starbucks or at home. The smartphone internally bridges the Wi-Fi connection to its cellular Internet data connection.

10.5 Smartphones

Of course, all cellphones are also computers. The keypad on a very basic phone can be used as an input device and the screen on the cellphone used as the display.

The keypad could be the regular telephone keypad, requiring the user to press the 2 button three times to select the character "c," for example. The keypad could also be a second qwerty-type keypad included with the phone like on a classic Blackberry.

Of course, the keypad and display can be implemented on a touch screen with an underlying graphic image of keys. Phones with touchscreens are called *smartphones*.

10.6 Data Plans

One must be very careful with billing plans when using a cellphone for Internet access. There can be one set of rates for voice, one set of rates for using the cellphone as a tethered modem or Wi-Fi access point, and a different set of rates for using a browser integrated in the cellphone.

If a user does not add a data plan to their account before using the phone to access the internet, they will be charged the "default," or "casual use" rate, which can be astronomically high, for example, $5 per MB.

$5 per MB is $5,000 per GB: $22,500 to download a regular DVD; $125,000 to download the data on a single-layer Blu-ray DVD.

It is not unusual to hear of people getting a bill of $20,000 for using their cellphone for internet access without a plan then watching YouTube and Netflix video.

One story had Canadian astronaut Chris Hadfield forgetting to sign up for a data roaming plan before spending a year orbiting the Earth on the International Space Station, and receiving a roaming bill for over a million dollars upon his return.

10.7 Converged Communications + Converged Device Achieved

The connection to the Internet supports phone calls, television, movies, music, email, web surfing and everything else, on the same connection on the same device: The Holy Grail of *convergence* people have been talking about for the last 50 years. Steve Jobs achieved it with his iPhone.

11 Mobile Network Operators, MVNOs & Roaming

11.1 Mobile Network Operator

Mobile Network Operator (MNO) is the term usually used to refer to a *facilities-based carrier*, i.e., a company that owns base stations, a mobile switch, backhaul between them, and spectrum licenses, and sells services to the public… and to other carriers.

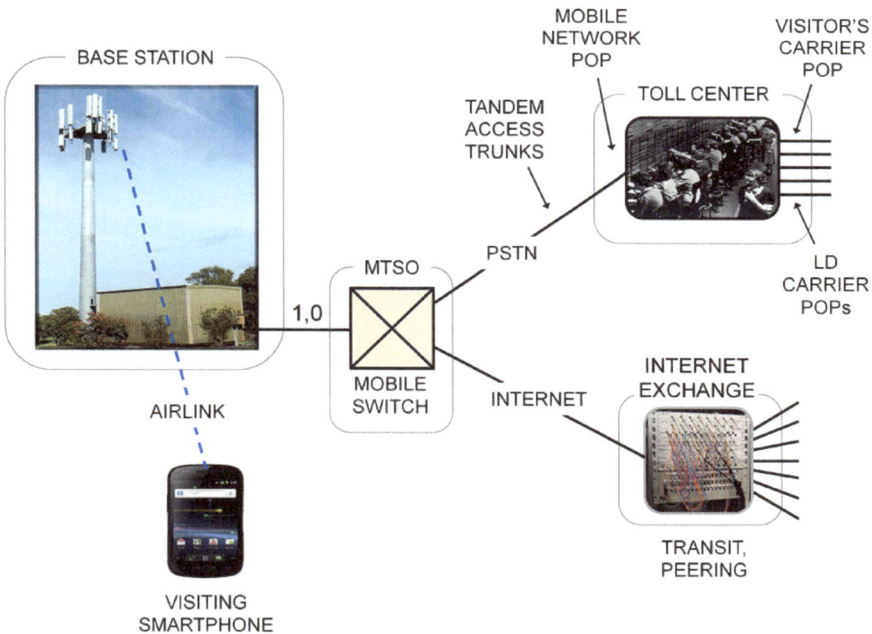

FIGURE 21 MOBILE NETWORK OPERATORS, MVNOS AND ROAMING

The MNO implements external links to other carriers for PSTN phone calls and for Internet traffic.

For PSTN phone calls, the MNO implements a fiber optic connection to a building traditionally called a Toll Center or Class 4 switching office. The termination of their fiber in that building is called a POP. It is their physical point of presence in the building.

Many other carriers have POPs in the building, including the Incumbent Local Exchange Carrier (ILEC), i.e., the local telephone company, Inter-Exchange Carriers (IXCs), i.e., long-distance providers, CATV companies, other mobile carriers, and any other company that wants to connect phone calls to a phone on the MNO's network.

The operator of the toll center, usually the ILEC, provides a switch in the Toll Center to switch phone calls from one carrier's POP to a different carrier's POP.

For Internet access, the MNO implements a fiber optic connection to one or more Internet Exchange buildings, where they pay the operator of the IX to route packets to other carriers with whom the MNO has established IP packet transit and peering arrangements.

11.2 Mobile Virtual Network Operator

Mobile Virtual Network Operator (MVNO) is the term used to refer to a non-facilities-based carrier... one that does not own the hardware or spectrum licenses or POPs. Instead, the MVNO enters into a long-term contract with one or more facilities-based carriers to have them supply a "white label" service that the MVNO sells.

Typically, the MVNO will develop a unique branding and sell smartphones and tablets to go along with its service.

When the MVNO deals exclusively with one carrier, the MVNO bill to the customer would be typically generated by the facilities-based carrier as a white-label service. If the MVNO is very large and deals with multiple carriers, the MVNO may operate their own billing system, which is a significant investment.

The facilities-based carrier bill to the MVNO includes a volume-discount rate for IP addresses and Internet traffic, voice-minute airtime and switched access to the POP for PSTN phone calls. The MVNO also has to pay for connectivity from the POP to other toll centers for "long-distance" connections, and the switched-access charge at the far end.

The rate plan the MVNO pays could be a mix of fixed-rate leases and usage-based billing.

Unless the MNO is obliged to sell capacity to MVNOs through regulations and tariffs, the nature of the plan is confidential business information.

11.3 Roaming

Roaming service is very similar to the service provided to MVNOs, in that it is the MNO that is providing the airlink, base stations, backhaul, mobile switch and connections to the PSTN and Internet.

Except that in the case of roaming, the visitor's phone has been activated on a different carrier's network, meaning they do not have a billing agreement directly with the carrier they are trying to connect to.

The roamer's carrier must have a roaming agreement in place with the carrier the roamer is connecting to before any roaming can happen.

When a roamer requests service, they will transmit the IMSI user ID to the carrier.

The first digits of the IMSI identify their country and their carrier. This is used to query the roamer's carrier whether they agree to pay the charges for the subscriber's roaming.

The IMSI is also used to retrieve information about the roamer from the roamer's carrier's switch location register once roaming is authorized. This is stored in a visitor location register in the roaming carrier's switch.

Once enabled and registered on the network, the visitor and their communications are treated the same as those of the home customers.

The difference is where the bill for the roamer's usage goes: the determinations the billing system makes from call records obtained from the switch, to direct charges to an individual customer's bill, or to a carrier's bill.

The billing between the two carriers could be any combination of per-minute or per-byte, volume discounts or fixed-rate plans.

In most cases, the carriers have customers roaming on each other's networks. Traditionally, the carriers would compare totals each month, and the carrier with the lower traffic would pay the other carrier a *settlement* for the difference.

In some cases, the carriers may have historically had generally equal traffic on each other's networks, and so to save significant administrative costs, come to an agreement to have no settlements between them: reciprocal free roaming.

The roamer of course pays for all of this in the end. The roamer's carrier bills the roamer flat-rate monthly, flat-rate daily, for a plan with a reasonable charge for voice minutes and mobile Internet, or the "occasional use rate", i.e., the no-plan astronomically high rate.

Roaming is an important feature for smaller players: they are facilities-based in selected cities, but to offer a national and international service to their customers, they must have roaming agreements in place with MNOs in other locations.

By denying roaming service to smaller or startup carriers, or charging an exorbitant price for roaming, an incumbent carrier can erect a barrier against competition. In many countries, the right to roam and the wholesale cost of roaming is regulated to encourage competition.

12 TDMA (IS-136) Time-Division Multiple Access

With the high-level picture in place, we now examine radio technologies used in 2G, 3G, 4G and 5G.

The 2G TDMA technologies in this lesson and the next are no longer in service. The lessons are included here so that you might know what GSM actually was, as part of a complete knowledge of wireless.

There are no longer any questions on IS-136 or GSM in the exam, so you may skip this lesson and the next and proceed directly to Lesson 14 which begins the discussion of CDMA if desired.

12.1 TDMA

One strategy used for spectrum sharing for the second generation of cellular (2G) is a combination of frequency and time multiplexing called Time Division Multiple Access (TDMA).

As illustrated in Figure 22, the frequency allotment is divided into radio channels, and a system of synchronized time slots on each radio channel is established. This allows several users to share the channel, interspersing their transmissions, and so increasing the capacity of the system.

Each user is assigned a time slot on an uplink radio channel, when they are allowed to transmit digitized speech or data, and a time slot on a downlink radio channel when they are allowed to receive digitized speech or data.

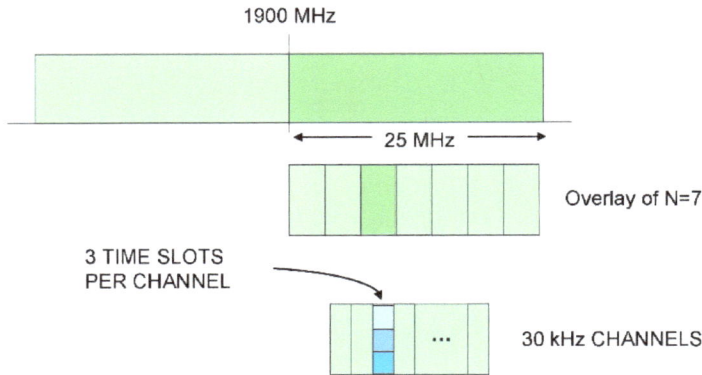

FIGURE 22 TDMA: TIME-SHARING RADIO CHANNELS

The users share a single channel in time, and so this is both a frequency division and time division strategy. This increases the capacity by three times.

12.2 IS-136 and D-AMPS

The company known today as AT&T Wireless in the US, and Rogers in Canada deployed TDMA conforming to the IS-136 and IS-54 standards.

This is sometimes called Digital AMPS or D-AMPS, as it used the same radio channels as AMPS, but with modems to carry digits, and three time slots per 30 kHz channel, within 1850-1920 MHz for the uplink and 1930-1990 MHz for the downlink.

12.3 Capacity Increase

Two channels (uplink and downlink) with three time slots each allows three simultaneous users per channel. A subsequent enhancement used voice compression to support six time slots per channel and so increasing the capacity by a factor of 6 compared to AMPS.

12.4 Inefficiency

However, time slots are not a very good way to increase the system's capacity, since a fixed amount of capacity is reserved for each user, whether they have anything to transmit or not.

Since speech is about 40% sound and 60% silence, reserving a fixed amount of capacity 100% of the time is inefficient.

For "data," such as sending and receiving email, or downloading web pages, the statistics are much worse.

The majority of time during data communication, nothing is being communicated – yet a TDMA system implements "circuit-switched data," i.e., reserves time slots 100% of the time during a data communication session.

13 TDMA (GSM)

The Global System for Mobile Communications (GSM), a 2G standard from the European Telecommunications Standards Institute (ETSI), is an implementation of TDMA that became, by far, the most popular mobile wireless technology deployed outside of North America.

13.1 Spectrum-Sharing Method

GSM defines 200 kHz-wide channels and eight time slots per channel. For each phone call, one time slot on one channel is reserved for uplink voice and one time slot on one channel is used for downlink voice.

A timeslot on each channel is used for timing and control information, yielding seven phone calls per two 200-kHz channels.

For those who like details, 270,833 signals per second are transmitted and the modulation scheme is either Gaussian Minimum Shift Keying (GMSK) or 8 Phase Shift Keying (8-PSK).

13.2 Inefficiency

A simple calculation will show that GSM supports fewer phone calls per Hertz than IS-136 TDMA. GSM also suffers from the same inefficiency as IS-136, reserving time slots 100% of the time during a connection even though speech happens only about 40% of the time.

13.3 Data, GPRS and EDGE

Data communications was treated the same way as voice, reserving time slots for the duration of a "connection" to a data server.

This was called *circuit-switched data*, and allowed 9.6 kb/s or 14.4 kb/s sessions. "High-speed circuit-switched data" allowed up to 43 kb/s using multiple time slots.

Subsequent add-ons to improve data communications included General Packet Radio Service (GPRS) allowing in practice about 40 kb/s packet data (less than obsolete landline dial-up modems) and EDGE (Enhanced Data for Global Evolution) at somewhat higher bit rates in practice.

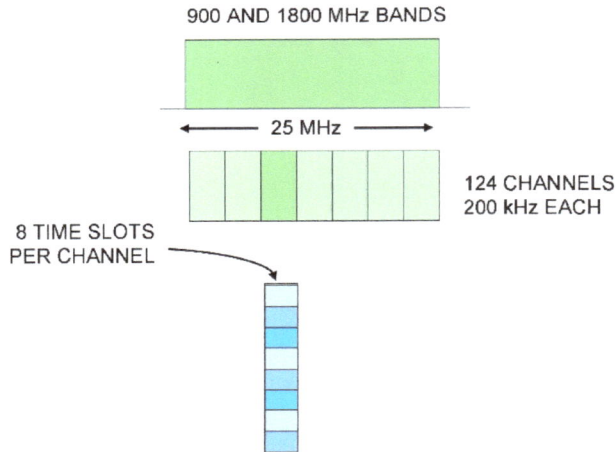

900 AND 1800 MHz BANDS

◄——— 25 MHz ———►

124 CHANNELS
200 kHz EACH

8 TIME SLOTS
PER CHANNEL

FIGURE 23 GSM TDMA

GSM supported hundreds of millions if not billions of users making phone calls on basic handsets, but did not support data communications to the extent required for web surfing or watching video.

13.4 Terminology: Misuse of the Term "GSM Phone"

It should be noted that the term GSM is sometimes misused to refer to all technologies employing TDMA, and/or all 2G and 3G technologies used by carriers that initially deployed GSM for 2G.

As we will see in upcoming lessons, in both the second and third generation, there were two incompatible spectrum-sharing methods, so it was necessary to identify which scheme a particular phone supported, particularly when it was being sold by a third party, not the carrier.

Some carriers and manufacturers incorrectly referred to a 2G phone that implemented IS-136 TDMA (AT&T Wireless and Rogers) as a "GSM phone", to contrast it to a 2G "CDMA phone".

While both use TDMA, GSM and IS-136 are not the same, making it an error to call an IS-136 phone a GSM phone.

In the third generation, CDMA phones that followed the 3G standard UMTS, favored by the companies that implemented TDMA for 2G, were also erroneously called GSM phones, to distinguish them from those that followed the competing 3G standard 1X.

Since 2G GSM TDMA and 3G UMTS CDMA are completely different, calling a 3G phone a GSM phone is an even more egregious terminology error.

14 CDMA Code-Division Multiple Access

Code Division Multiple Access (CDMA) was deployed for the second generation by Verizon and Sprint in the US, and Bell and TELUS in Canada; it was also used for all serious 3G technologies worldwide.

3G CDMA is still in service. Like everything else, CDMA systems will be turned off and the spectrum re-used for newer LTE and 5G technologies using a different scheme called OFDMA, covered in an upcoming lesson.

Understanding CDMA remains part of the core knowledge set in wireless.

CDMA is a completely different strategy for spectrum sharing than FDMA or TDMA.

14.1 Carriers

In a public mobile 2G CDMA system, the frequency allocation is divided into 1.25 MHz-wide bands called *carriers*, illustrated in Figure 24.

One carrier is used for the downlink, or forward direction (base station to handset), and a second is used for the uplink, or reverse direction (handset to base station).

14.2 Codes

In a CDMA system, users transmit at the same time, in the same place, on the same carrier.

User data is distinguished by transmitting a unique pattern of bits called a *code* to represent the users' 1s and 0s.

The codes are designed so that if multiple users' codes are transmitted at the same time, it is possible to analyze the received total to determine which codes were transmitted and which were not.

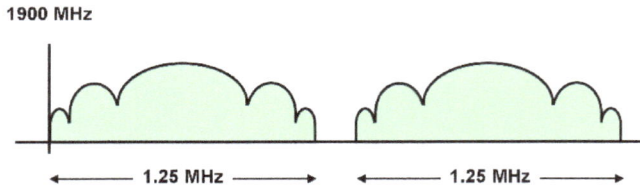

FIGURE 24 CDMA CARRIERS

This is analogous to being at a cocktail party where everyone is speaking at the same time in the same space at the same frequencies, but each pair of people are speaking languages with unique words – and you only understand the words in your partner's language.

You can understand what your partner is saying by listening to the hubbub of words you hear and correlating them to your vocabulary. Only your partner's words make sense, everything else sounds like noise.

A *code* is a typically 64-bit-long string of 1s and 0s. A code is established for each user, for each direction. To indicate a "0," the code is transmitted. To indicate a "1," the binary complement of the code is sent.

14.3 Forward Error Correction

Only hundreds of the possible 2^{64} = 18 billion billion codes are used. The valid codes are spread out evenly between 0 and $2^{64} - 1$, so that if some bits are received in error, the error can be corrected by choosing the closest valid code.

This is called error-correcting coding or Forward Error Correction (FEC), implemented in many radio systems to mitigate errors due to fading.

14.4 Variable-Rate Coding

CDMA can employ variable-rate voice coding, which means the number of bits per second varies according to the complexity of the sound.

When there is silence, which is 60% of the time during a phone call, the bit rate is very low.

Compare this to TDMA, where the bit rate is constant, even when there is silence to be transmitted.

14.5 Packetized Voice and Data

CDMA systems package both voice and data into packets for transmission. CDMA is more spectrally-efficient than TDMA or FDMA, affording the greatest number of phone calls per Hertz of radio band. Since packets are transmitted, it is also efficient for data communications.

14.6 Qualcomm, IS-95A and IS-95B

2G CDMA systems followed the standard IS-95, branded cdmaOne by Qualcomm, a company that patented power control and synchronization methods used in CDMA.

IS-95A allowed up to 14.4 kb/s of data communication. IS-95B allowed up to 115 kb/s of data, at the expense of fewer voice calls in the same 1.25 MHz carrier.

15 Spread Spectrum

CDMA is a *spread spectrum* technique.

This lesson covers the meaning of spread spectrum and how it is accomplished with CDMA, compared to other schemes like frequency-hopping used in Bluetooth.

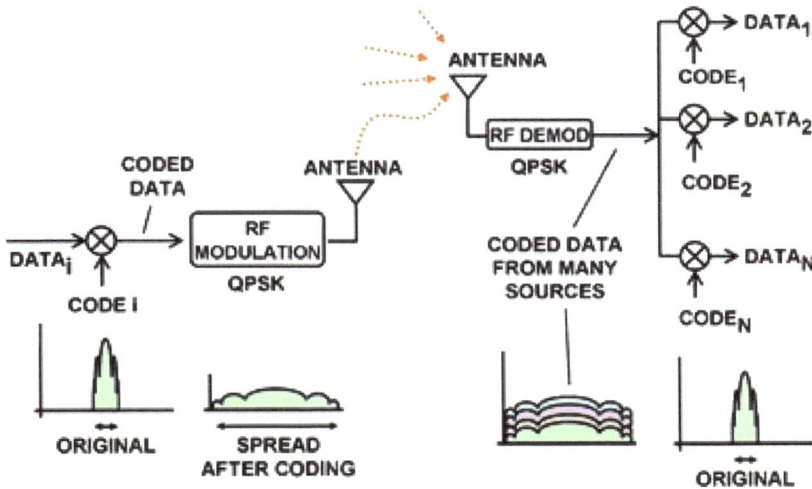

FIGURE 25 TRANSMITTING CODES INSTEAD OF SINGLE BITS SPREADS THE ENERGY ACROSS A WIDER FREQUENCY BAND

A CDMA system transmits a code to represent a user's 1 or 0. Codes are typically 64-bit-long sequences of 1s and 0s.

15.1 Chips and Chipping Rate

To distinguish between the bits in the users' data and the bits in the resulting stream of codes, which has 64 times as many bits as the user data, sometimes the term *chip* is used to refer to a bit in the coded bit stream.

In this terminology, the code is said to have 64 chips (instead of 64 bits), and the *chipping rate* is the rate at which the coded bit stream is transmitted.

The number of bits per second being transmitted is the chipping rate, which is higher than the user data rate by a factor of 64.

15.2 Spreading

Modulation techniques including Quadrature Phase Shift Keying (QPSK) are used to represent these bits or chips on the airlink.

Information theory, in a relationship discovered by Claude Shannon at MIT, tells us that the number of bits per second is proportional to the width of the frequency band – and vice-versa.

Since the number of bits per second transmitted is the chipping rate, 64 times higher than the user data rate, this means that the energy produced by the radio is spread across a proportionally wider frequency band or *spectrum* compared to just transmitting the user 1s and 0s.

The term *spread spectrum* is used to refer to this phenomenon.

15.3 Direct Sequence vs. Frequency Hopping Spread Spectrum

Public mobile CDMA communication systems use Direct Sequence, where the coded bit stream is transmitted at a single carrier frequency.

Other systems, including Bluetooth, use frequency-hopping to effect the frequency spreading, where the coded bit stream is transmitted at a carrier frequency that frequently changes. This is not used in mobile cellular.

As illustrated in Figure 25, the waveform received at the base station is the aggregate of all the handset transmissions, their energy added together.

The receiver demodulates the QPSK to extract the 1s and 0s that are the codes added together, then performs a mathematical function known as *convolution* in the time domain, *correlation* in the frequency domain, to detect which codes it has received.

The receiver then knows if a particular handset transmitted, and, if so, whether it was transmitting a 1, or a 0 which was digitized speech or data.

15.4 Error Correction

Only a few of the possible 18 billion billion codes implied by 64-bit-long codes are actually assigned to phones and transmitted. The valid codes are spread out evenly in the range from 0 to 2^{64} -1, and the codes between are not valid.

This allows the receiver to perform error correction: if an invalid code is received, the receiver makes a decision that the nearest valid code was the one actually transmitted, correcting the error.

15.5 Rake Filters and Multipath

The receiver performs this mathematical operation, then performs it again a short time later, then again a bit later, then again.

This is called a *rake receiver,* and has the effect of detecting the code via multiple bounce paths, and the results used to reinforce the final answer.

In contrast to other transmission schemes, where multiple copies of the signal arriving at different times due to following different bounce paths with different lengths interfere with the detection process, in a CDMA system, each copy of the signal reinforces the decision-making.

This is happening in both directions at the same time, on separate 1.25-MHz carriers.

16 CDMA Operation and Patents

Deploying CDMA in a mobile, multi-user, multi-base-station network is an extremely complex proposition, involving communication with multiple base stations at the same time, which requires tight synchronization and extensive signal processing.

16.1 Communication via Multiple Base Stations

The base stations in adjacent cells or adjacent sectors typically communicate on the same carrier, causing transmissions from the terminal to be received at multiple base stations. This allows the switch to communicate to the mobile through multiple base stations simultaneously, and is used to improve the handoff performance and bit error rate.

At any given time, the mobile will be receiving a transmission from its primary base station (the one with the strongest carrier), which may be added together with transmissions from one or more secondary base stations with weaker carriers.

16.2 Multipath

On the downlink, the signal received by the mobile from the primary base station will be a composite of energy following many physical paths: direct transmission, reflections off objects like buildings and tree leaves, and energy refracted around obstacles; this is called *multipath*.

The copy of the signal following each path will arrive at the mobile at a slightly different time, since each path has a different length. Signal processing in the receiver accounts for these timing variations before adding the signal from each path together to produce a final result.

The switch can also transmit the user data from secondary base stations. Each of these carriers will arrive at a slightly different time at the mobile, since they are physically different distances away.

These signals are treated like multipath reception from a single base station and improve the accuracy of the result.

On the uplink, the data received by the primary base station and the secondaries is compared and the most reliable is chosen, improving the error rate.

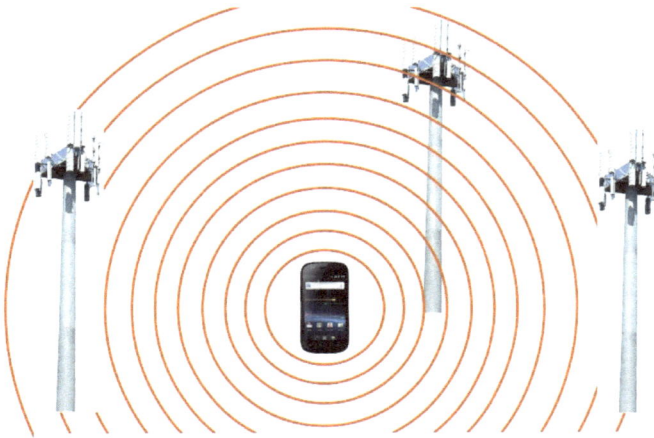

FIGURE 26 SOFT HANDOFFS

16.3 Soft Handoffs

If the mobile moves too far away from the primary base station, its carrier will no longer be the strongest, and so one of the base stations that was a secondary becomes the primary, and the former primary becomes a secondary.

This is called a *soft handoff* and involves no loss of communications during the handoff.

This is in contrast to AMPS, TDMA and other technologies where communication with the primary base station is terminated then re-established with a new base station, causing a temporary interruption or muting of the communications.

16.4 Walsh Codes and Pseudonoise

The system is different in the downlink and uplink directions. On the downlink, the base station transceiver transmits all handsets' codes, called Walsh codes, at the same time.

For the uplink, the mobile handsets cannot be synchronized so that their transmissions are received at the base station at the same time, so the handset transmits a code derived from its serial number and an almost-random pattern of bits called a Pseudo-Noise (PN) sequence.

16.5 Base Station Identification, Short Codes and Timing

To identify itself, each base station transmits a repeating pattern of bits, sometimes called the *short code,* on a control channel. Each base station transmits the same code, but uniquely offset in time or *phase* compared to the other base stations. The phase identifies the base station.

For this to work, the timing of transmission of the bits in this code must be very tightly synchronized between base stations. A very precise clock derived from the Global Positioning System (GPS) is used to synchronize base stations.

16.6 Power Control

The power received at the primary base station must be the same from all handsets – regardless of how far the handset is from the base station.

To achieve this, the broadcast power of the mobile is adjusted as it moves: the power is increased as the mobile moves away from the primary base station and decreased as it moves towards the primary base station. IS-95 systems adjust the power transmitted by the mobile 800 times per second.

16.7 Qualcomm

The company Qualcomm patented certain methods of power control and synchronization necessary to make this work, and sells chips implementing these patented techniques, or collects patent royalties from manufacturers who make their own chips implementing the techniques.

Various interested parties attempted to develop methods of power control and synchronization that do not use Qualcomm's patented techniques, but were unsuccessful.

It is estimated that the cost of every CDMA phone included approximately $35 in patent royalties, a large portion of which go to Qualcomm.

Even though new phones today support LTE and 5G, both of which use OFDM, a technique completely different than CDMA, these phones are typically backward-compatible with 3G for those areas not yet served by 4G… meaning that the CDMA royalty payment to Qualcomm continues.

17 3G: CDMA 1X, UMTS and HSPA

The third generation of cellular is usually referred to as 3G. The main objectives of the third generation were to improve capacity, and to increase the number of bits per second that could be transmitted over the airlink for mobile wireless high-speed Internet access and video.

3G will disappear, replaced with 4G and 5G. At time of writing, an installed base of 3G is still in service.

17.1 IMT-2000

To try to avoid a repeat of the 2G CDMA vs. TDMA dichotomy, in 2000, a standards committee attempted to define a world standard for 3G called IMT-2000.

They failed.

The result was a "standard" describing five incompatible implementation variations. Like many other technologies, we ended up with one solution for "North America" and a different solution for "Europe".

The two serious variations in IMT2000 both specified CDMA as the method for spectrum-sharing, but ultimately disagreed on the width of the radio bands and how many bands there should be.

17.2 1X

Service providers using CDMA for 2G, primarily in North America and certain Asian countries, favored a strategy that was a software upgrade from 2G, employing existing 1.25 MHz radio carriers and allowing multiple carriers.

This is called IMT-MC or CDMA multi-carrier, branded CDMA2000 by Qualcomm. A single 1.25-MHz carrier version of this is referred to as "1X".

In the 1X systems, power control and synchronization were accomplished using techniques patented by Qualcomm, meaning a royalty to Qualcomm for every cell phone and every base station transceiver, and the use of the United States Government's Global Positioning System.

FIGURE 27 1X AND UMTS

Service providers using GSM TDMA for second generation, primarily cellular carriers outside North America, favored the deployment of CDMA in a 5 MHz wide band. This was called, IMT-DS, Direct Spread, Wideband CDMA (W-CDMA) and Universal Mobile Telephone Service (UMTS).

European operators did not favor paying American royalties, nor a network using American GPS.

They embarked on a seven-year-long series of fiascoes attempting to implement CDMA on 5 MHz carriers, circumvent Qualcomm patents, and avoid GPS.

In early trials, the processor on the phone drew so much current from the battery it got hot enough to burn users' hands. After several other trials failed, a Euro-GPS called "Galileo" had to be created for UMTS.

This delayed deployment of 3G in Europe until 2007; 1X was deployed and working years earlier.

The tipping point was reached in the summer of 2007, when more **new** activations on these carriers' networks were 3G CDMA (UMTS) instead of 2G TDMA (GSM). The 2G TDMA technology GSM still had far more users, but like 1G analog, GSM has now disappeared.

17.3 Data-Optimized Carriers

For Internet access and watching video on cellphones, variations optimized for "data" were deployed by both camps on carriers separate from those used for telephone calls.

The 1X camp developed a variation called 1X Evolution Data-Optimized (1XEV-DO), allocating a 1.25-MHz carrier for data communications and promising 2.4 Mb/s heading up to 70 Mb/s.

The UMTS camp developed High Speed Packet Access (HSPA), promising 14.4 Mb/s then 42 Mb/s, then Evolved HSPA+ promising 168 Mb/s down and 22 Mb/s up under perfect conditions.

17.4 Capitulation

Market forces finally pushed the two camps together. The fact that there were far more 2G GSM users meant that GSM phones were less expensive and had better features. This trend was continuing into 3G, where UMTS phones would have the same advantage over 1X phones.

Another fact was that Steve Jobs only permitted carriers operating GSM or TDMA systems to have the iPhone, then only those with HSPA systems to have the iPhone 3G. As the iPhone was at the time the most popular consumer electronics device, this severely tilted the playing field.

Finally, the 1X camp threw in the towel and decided to go with the UMTS camp's proposal for the fourth generation to level the playing field.

As soon as that decision was made, the deployment of 1XEV-DO was capped, and many 1X carriers including Verizon in the US and Bell and TELUS in Canada began deploying HSPA then HSPA+ instead.

As soon as those carriers completed their HSPA networks, Steve Jobs allowed the iPhone on their networks.

One of the legacies of Steve Jobs will not just be the iPhone, but ending the mobile radio standards wars.

18 4G LTE: Mobile Broadband

18.1 Introduction

In this lesson, we'll explore the technology that emerged as the consensus for 4G: the fourth generation of mobile cellular radio communications.

The radio technology employed is called the Universal Terrestrial Radio Access (UTRA), and the network is the Universal Terrestrial Radio Access Network (UTRAN) Long-Term Evolution (LTE).

An upgrade of LTE supporting higher bit rates called LTE-Advanced and "E-UTRA / E-UTRAN stage 2" was released after LTE.

Many people refer to all of the above as "LTE".

First, we'll provide an overview of LTE and its characteristics.

In the second part of this lesson, we'll take a deeper dive into OFDM and how it is implemented for LTE.

This is a relatively technical discussion explaining what subcarriers are and why they are used, revisiting modulation, QAM, QAM-64 and baud rate, then why the baud rate is equal to the subcarrier spacing for OFDM and what that accomplishes.

Then OFDMA, how the base station communicates to the handset and vice-versa, and how the actual radio waveform broadcast by the base station is calculated as a function of time.

In the third part of this lesson, we'll review the four LTE standards releases and the features each brought, including those annoying Amber Alert and Presidential messages.

18.2 LTE for the UTRAN

After more than 20 years of incompatible 1G, 2G and 3G systems, a universally-accepted standard for mobile radio was achieved with the fourth generation. By inventing the world's most popular consumer electronics device – the iPhone – and only allowing carriers following the "European" standards, often called "GSM" carriers, to sell it, Steve Jobs almost single-handedly brought the cellular standards war to an end.

To get the iPhone, North American service providers agreed to go along with the "European" standards for 4G; in particular, the Third Generation Partnership Project (3GPP) standards group Release 8, Universal Terrestrial Radio Access Network (UTRAN) Long Term Evolution (LTE).

LTE is an all-IP network, moving VoIP and Internet traffic over the airlink. LTE's spectrum-sharing method, Orthogonal Frequency Division Multiplexing (OFDM), is a return to first-generation Frequency Division Multiple Access with some major improvements.

The 3GPP Technical Report 25.913 contains the detailed requirements specification for LTE.

The system architecture, in Technical Specifications 36.300 and 36.401, is simplified to two principal network elements: evolved Network Base stations (eNBs) and Evolved Packet Cores (EPCs). eNBs communicate with EPCs, with each other and with user equipment as illustrated in Figure 28.

Another improvement is the definition of a thousand or more channels, called *subcarriers* in OFDM, and dynamic assignment of multiple channels to users for parallel downloads and efficient implementation of high bit rates. On each channel is a modem running QAM or QPSK.

A device called the Radio Resource Controller (RRC) at the base station manages the assignment of channels, and also controls the wake/sleep state of the phone to conserve battery life.

Multiple-Input, Multiple-Output (MIMO) antenna designs can increase the bitrate using spatial multiplexing, which is basically gluing several transceivers and antennas together and communicating in parallel.

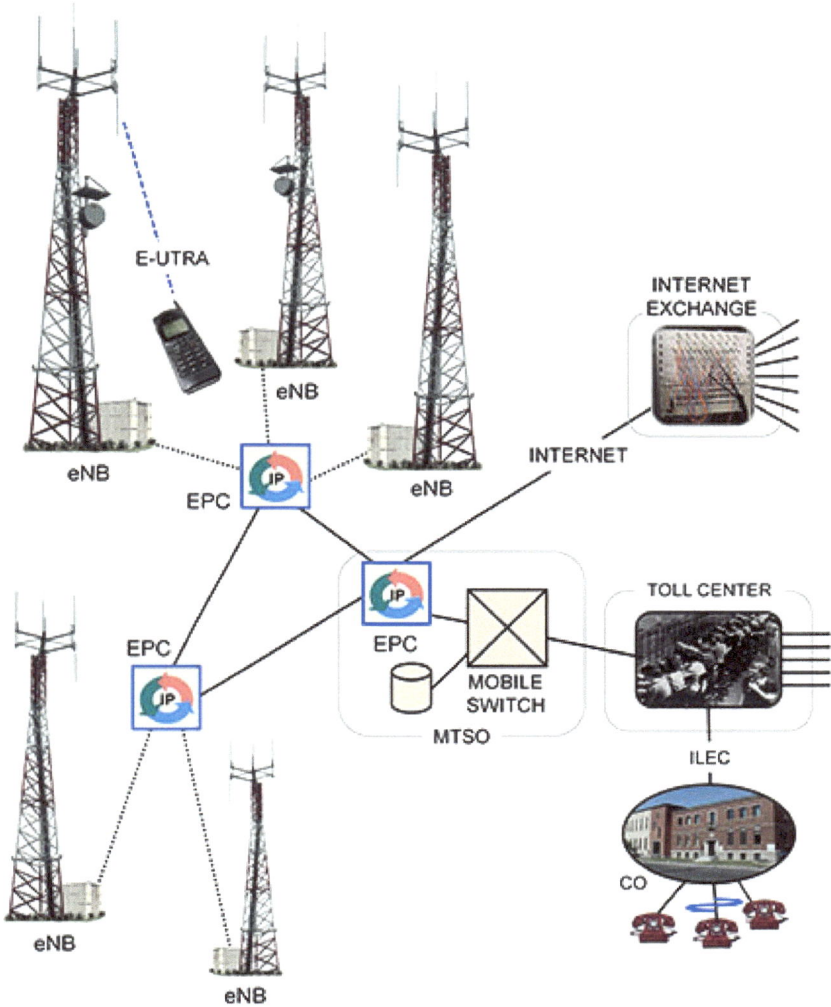

FIGURE 28 LTE SYSTEM ARCHITECTURE

Service providers can deploy LTE on radio carriers 1.4, 3, 5, 10 or 20 MHz wide. The highest bitrates, on the order of 100 Mb/s are achieved on a 20 MHz carrier. 5 MHz carriers allow overlay on frequency allocations for UMTS.

Cell sizes can range from femtocells measured in the tens of meters (in-home), picocells up to 200 m (office building), microcells between 200 m and 2 km, which includes cells in urban areas typically 1 km, cells in suburban areas typically 5 km, and macro cells in rural areas 30 km and theoretically up to 100 km in diameter.

18.3 Modems, Modulation, and How OFDM Moves 6-Bit Numbers Simultaneously to Different People on the Same Carrier

The modulation and spectrum-sharing scheme for LTE is OFDM, which is different than FDMA, TDMA and CDMA.

OFDM is used in LTE, 5G, Wi-Fi, DSL and Cable modems. In this part, we'll take a deeper dive into OFDM and its implementation for LTE. This is a relatively technical discussion; we'll try to hit the main ideas.

The basic idea with OFDM is the definition of hundreds or thousands of subcarriers within the main carrier. A subcarrier is a single frequency which will be modulated like any other modem carrier signal, spreading energy in a small band around the subcarrier.

Information is transmitted on the subcarriers in parallel, implementing not only a high total bitrate, but also the ability to communicate on a particular subcarrier to a particular device in a broadcast situation. These were called "channels" in 1G.

18.3.1 Modulation

To recap the discussion from Course 2231: *modulation* means producing a voltage that is vibrating at a single pure frequency, called a *carrier frequency*, and to communicate bits, changing aspects of it in discrete steps. One aspect is the volume or *amplitude*. Changing the volume up and down makes changes that represent bits.

Another aspect is the *phase* of the vibration: when the peak of the cycle is happening in time with respect to others. Changing the time of the peak so it happens a bit earlier than others, or making it happen a bit later is making changes that can represent bits as illustrated in Figure 29.

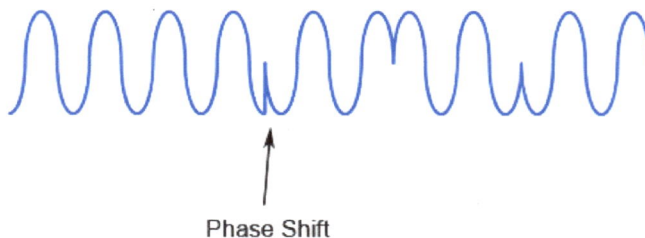

Phase Shift

FIGURE 29 QUADRATURE PHASE SHIFT KEYING

Combinations of phase and amplitude shifting is called Quadrature Amplitude Modulation (QAM). QAM-64 means 64 possible different combinations of 8 different phases and 8 different amplitudes. Each combination, also called a symbol or signal, is assigned a number. Binary requires 6 bits to give binary numbers to each of the 64 combinations.

18.3.2 Communicating Six Bits: Sending One of 64 QAM Signals

To communicate six bits in one fell swoop on a subcarrier, the transmitter generates electricity vibrating at the subcarrier frequency with the phase and amplitude corresponding to the combination indicated by the six-bit number.

The electricity is turned into radio by antennas for communication through the air.

When the receiver detects electricity at that single pure carrier frequency, it measures the phase and amplitude, and once it has decided, outputs the six-bit number of the combination it is hearing, and Bob's your uncle.

18.3.3 Baud Rate Equal to Subcarrier Spacing

To get many bits per second, the procedure has to be repeated often! Repeat it once per second, that's 6 bits per second; the combination of phase and amplitude of the carrier is maintained for one second. The rate at which the procedure is repeated is called the baud rate, signaling rate and symbol rate.

The baud rate, how often a new combination can be applied to the carrier to communicate another 6 bits, is limited by interference called harmonics, where energy gets spread into adjacent channels, and interferes with communications on other channels.

When the baud rate is the same as the subcarrier spacing, the harmonics from other subcarriers cancel out. Eliminating this source of interference allows successful data transmission in parallel on closely spaced subcarriers.

This is a prime design characteristic of OFDM, and is the sweet spot for baud rate in terms of efficiency. Increasing the baud rate to get even more bits per second would require spacing the channels further apart to avoid harmonics, and would end up communicating fewer bits in a given carrier.

18.3.4 LTE Specification and OFDMA

To support many users in a broadcast system, many subcarriers are used, spaced at regular intervals across a 20 or 100-MHz carrier (Figure 30). The Radio Resource Controller allocates individual subcarriers to individual users, called Orthogonal Frequency-Division Multiple Access (OFDMA). It can also allocate all the subcarriers to one user to do a massive parallel download.

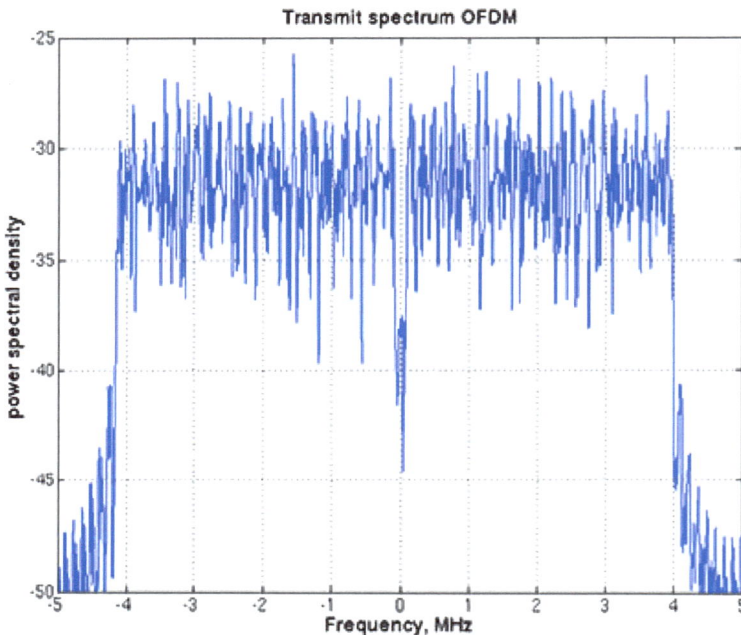

FIGURE 30 OFDMA - SUBCARRIERS ALLOCATED TO INDIVIDUAL USERS

In the LTE standard, the subcarriers are spaced 15 kHz apart, and the signaling rate is 15,000 baud. A 20-MHz carrier would allow up to 1,333 subcarriers; and if each were running a QAM-64 modem, six bits per baud, the total bit rate would be 120 Mb/s.

Prior to modulation, Forward Error Correction is implemented, adding redundancy to the bit stream so that correct decisions can be made based on maximum likelihood in the presence of impairments like noise and fading. The bit stream is also shuffled or interleaved, re-arranging the order of the bits in time so that burst errors are no longer sequential errors.

In this example, at the base station transmitter, bits to be transmitted to devices are broken into groups of six and transmitted using modems on subcarriers determined by the Radio Resource Controller.

Notionally, for each of the 1,333 subcarriers, the phase and amplitude shift corresponding to the six bits is determined, and the appropriate signal is generated. All of the subcarrier signals combined on one wire, energy evenly spaced across the 20-MHz carrier, and transmitted with no changes for 1/15,000th of a second.   This indicates the value of 7,998 bits (6 x 1333).

The electricity is turned into electro-magnetic waves by the base station antenna, which propagate to the handset at the speed of light. The handset's antenna turns the received radio energy across the 20 MHz carrier back to electricity on a wire in the handset.

The handset has 1/15,000th of a second to measure the signal on the subcarriers it has been assigned, decide what signal in the constellation is being transmitted, and pass the corresponding pattern of six 1s and 0s up the protocol stack. This process repeats 15,000 times per second.

In the uplink, an added DFT results in Single-Carrier FDMA, a signal transmitted on one subcarrier, in contrast to OFDMA which is a multi-carrier transmission scheme. This reduces the cost of the radio transmitter in the handset.

The values of the subcarriers are added together at the base station to produce the signal to be transmitted, first in the form of electricity, then converted to radio by antennas. This signal has a bandwidth of 20, 40 or 100 MHz, and is frequency-shifted to the spectrum the operator is leasing (Figure 31).

The voltage value of the signal to put on the wire leading to the antenna at any one instant is in fact calculated as a function of time in a single step with a digital signal processing operation called an Inverse Discrete Fourier Transform. The subcarriers are represented by a set of orthogonal polynomials, which could be thought of as separate dimensions, and the bits to be transmitted are mapped on to them in a rectangular matrix. Solving the matrix with a particular value for time yields the value of the signal at that particular instant.

The value of the signal must be calculated and adjusted more than twice as often as the frequency of the carrier.

subcarrier f_1

1

SUBCARRIERS ARE ADDED
TO PRODUCE WAVEFORM
TO TRANSMIT BY RADIO

subcarrier f_2

0

101 • • • 1

subcarrier f_3

1

Σ

• • •

subcarrier f_n

1

CALCULATED IN ONE STEP
WITH AN INVERSE DISCRETE
FOURIER TRANSFORM

BLOCK OF n BITS
BECOMES ON/OFF CONTROLS
FOR n SUBCARRIERS

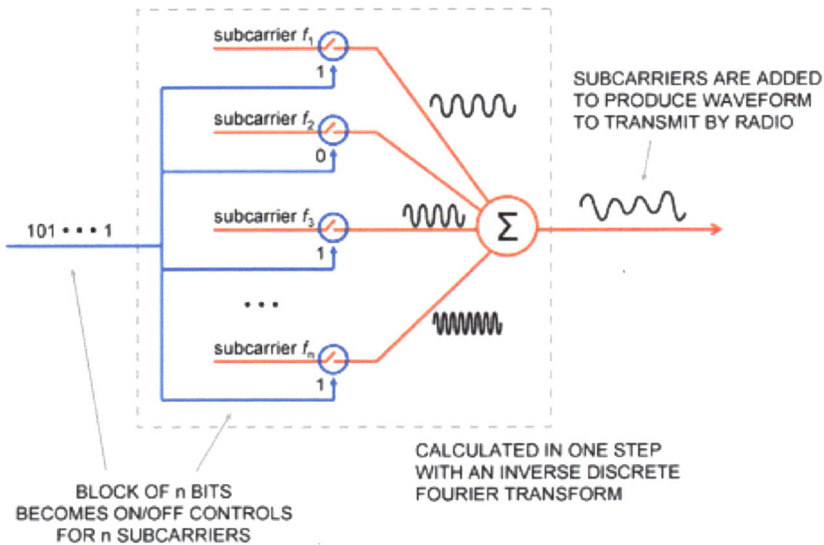

FIGURE 31 SIMPLE EXAMPLE OF OFDM WITH ASK OF SUBCARRIERS

18.4 3GPP Releases

3GPP Release 8 was the beginning of LTE, with the UTRAN and the 15-kHz OFDM air interface.

Release 9 was enhancements to LTE. This included the Public Warning System, which wakes you up at 2 in the morning with a horrible screeching sound to alert you about events happening 800 miles away. Not sure if that is an enhancement.

Release 9 also specified enhancements to device location, including in order of accuracy: GPS, propagation delay triangulation from base stations, and enhanced cell ID.

Release 10 defined LTE-Advanced, along with the E-UTRA and E-UTRAN. The biggest feature was increasing to 1 Gb/s download, which met the bitrate requirement specified by a committee at the ITU to qualify as "4G". (Everyone else called Release 8 "4G").

Release 10 also supported 40 MHz carriers, and carrier aggregation, where an operator could piece together smaller licensed bands to make up to 100 MHz carriers, improving the transition from 2G and 3G systems to LTE.

Releases 11 through 13 were enhancements to LTE-Advanced. In Release 12, a new LTE category 0 was introduced, for IoT devices needing little bandwidth and power conservation.

Other enhancements included various efficiency and network management enhancements preparing for massive machine communications for IoT, enhancements for small cells, control of Wi-Fi vs. cellular data plan usage by the handset, and device location enhancements.

18.4.1 The Eventual Pivot To 5G Across the Spectrum

Release 14 was the beginning of 5G and New Radio, which has a different air interface and is not compatible with LTE.

LTE will continue to be deployed and supported for years as operators add to installed base and leverage the cost of infrastructure investments.

At some point in the future, like AMPS, TDMA, GSM, EDGE, GPRS, 1X, 1XEV-DO, UMTS and HSPA before it, LTE will be considered too slow by consumers and marketing departments, and will disappear, to be replaced by the next generation, 5G across the spectrum.

An interesting side note: one of the reasons for the 2G and 3G standards wars was the requirement to pay American company Qualcomm royalties on patents for several techniques necessary for a mobile CDMA system. LTE is not CDMA, so those royalties are avoided... but it turns out that Qualcomm filed or has purchased many patents that underpin LTE and 5G.

Additionally, since 4G and 5G phones have to be backwards-compatible with installed-base 3G CDMA systems until those are retired, Qualcomm sees "no impact" on patent royalty revenue for the first ten years of LTE development according to COO Sanjay Jha. Qualcomm's royalties on 5G phones using their chips are 3.5% of the retail price of the phone.

19 Dynamic Assignment of Subcarriers

First-generation analog cellular, fourth-generation and fifth-generation digital cellular all use Frequency-Division Multiplexing, where the spectrum is divided into radio channels that are assigned to users.

In 4G and 5G systems, there are modems in the base station and in the mobile operating on each channel, and a user can be assigned many channels to implement a very high bit rate, transmitting data on the channels in parallel.

19.1 1G vs. 4G and 5G

In the first generation, there was one channel per user, and no modems built into the system.

In 4G LTE and 5G, the assignment of multiple radio channels to users is dynamic. As illustrated in Figure 32, at one instant, a user could be assigned dozens or even hundreds of channels for a burst download of a file or a video, then an instant later those channels would be assigned to another user.

The dynamic assignment of subcarriers to users is implemented by the Radio Resource Controller (RRC) in the base station.

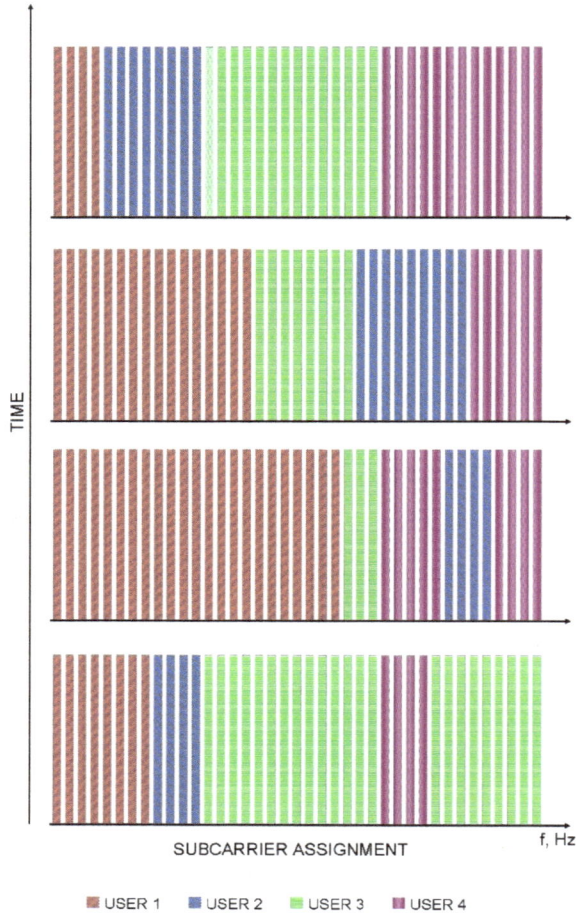

FIGURE 32 DYNAMIC ASSIGNMENT OF SUBCARRIERS

20 5G New Radio: Enhanced Mobile Broadband, IoT Communications

20.1 Introduction

In this lesson, we'll cover 5G, understanding the frequencies it will be deployed on and the bitrates that will be available, ranging from "not very many, once in a while" on lower frequencies for IoT devices, through "faster Internet" on conventional frequencies for your smartphone, to ultra-broadband for VR using millimeter-wave bands.

20.2 3GPP Release 15

Standards for 5G are developed by the increasingly inaccurately-named Third-Generation Partnership Project (3GPP), the "European" standards organization that brought us LTE as the worldwide standard for mobile communications.

3GPP standards are referred to as Releases, which are... released every year or two.

Notable releases are Release 8, which specified LTE, Release 10 LTE-Advanced, Release 13 which added elements for prioritizing mission-critical communications for public safety applications, and release 15, the first full standalone standard for 5G.

20.2.1 Immediate Impact Of 5G: More Bits Per Second

The immediate impact of 5G is more bits per second; the "speed" of the Internet on users' phones and cellular-to-Wi-Fi devices will increase another notch.

5G uses a new air interface called New Radio in the standards committees, which has 40% better spectral efficiency (bits/second per Hertz) than LTE.

5G also supports massive MIMO, allowing massive parallel communications again increasing the bit rate, and brings with it new radio bands, where the channel widths can be up to 100 MHz wide, compared to 20 and 40 MHz for LTE.

From a business point of view, one effect of this increase in available bits per second will be to make it easier for operators to offer unlimited data plans with new 5G devices.

20.3 New Spectrum

Spectrum for 5G is grouped into "above 6 GHz" and "below 6 GHz". Some of the new bands defined in North America are listed in Figure 33.

Band Name	Lower	Upper	Bandwidth	Channel Bandwidth
600 MHz	614	698	70	5
700 MHZ	698	806	108	1 - 12
800 MHz	806	894	88	5 - 10
2.5 GHz	2500	2690	190	18 - 50
3.5 GHZ	3550	3700	150	10 - 40
3.7 GHz	3700	4000	280	20 - 100
26 GHz	24250	27500	3250	120 - 400
28 GHz	26500	29500	3000	120 - 400
28 GHz US	27500	28350	850	120 - 400
37 GHz	37000	37600	600	120 - 400
Upper 37	37600	38600	1000	120 - 400
39 GHz	37000	40000	3000	100 - 1400

FIGURE 33 NEW SPECTRUM

5G will be deployed on existing spectrum, eventually replacing LTE.

Interest is currently focused on deployment of 5G on new frequency bands under 6 GHz, particularly the 2.5 GHz band which has a favorable balance of available bandwidth and signal propagation through obstacles.

The 3.5 GHz band has significant bandwidth available for sectorized systems, but does not propagate through obstacles like walls and trees.

Other new spectrum is at comparatively low frequencies, the 600-, 700- and 800-MHz bands. These bands have the advantage of long reach, i.e., signals can propagate over long distances and inside buildings; but the number of 20- and 40-MHz bands available for sectors is limited.

"Above 6 GHz" is where the real broadband will be provided.

20.3.1 Millimeter-wave

Millimeter-wave (mmWave) means radio frequencies where the wavelength is measured in millimeters: 30 – 300 GHz. With light propagating at 3×10^8 m/s in a vacuum, the length of the radio wave is 10 mm at 30 GHz, and shortens to 1 mm by 300 GHz.

At these very high frequencies, it is possible to have very wide carriers, channels measured in GHz wide – far beyond the 40 MHz of LTE – implementing ultra-broadband: hundreds of Gb/s.

However, at these very high frequencies, problems with penetration through the air itself limits the maximum useful airlink range to 150 meters with line of sight.

20.4 5G Design Goals and Use Cases

In the lead-up to 5G, to frame and guide the development of the standard, the 3GPP published a list of design goals and use cases.

Design goals included:
- The next level of broadband: 10 Gb/s peak downloading;
- Support for a very high number of connections: 1 million devices per square kilometer (2.6 million per square mile);
- Very low network round-trip delay or latency; and
- Ultra-reliable communications

Use cases illustrate the intended use of the new radio standards, and allow a comparison between what is written in the standards document, and what needs to be in the document.

20.4.1 Enhanced Mobile Broadband

The first use case is Enhanced Mobile Broadband: a significant increase in bit rates available to users. Using wider (2 GHz) bandwidths on new higher frequency spectrum, this use case is targeting to deliver 10 Gb/s of downlink throughput – 100x that of single-carrier LTE.

At 30 GHz, the maximum size of a cell is 300 meters in diameter, meaning the need for many antennas, so the application for ultra-broadband is urban areas, highways, in-building and other densely populated areas where the build cost of so many antennas and radios can be repaid.

Conventional spectrum, e.g., 2.3 and 2.5 GHz, allows cell sizes of 10 km (6 miles) or more, and so is also suitable for less densely-populated areas. Deploying 5G on these frequencies gets users 40% more bits per second than LTE.

20.4.2 Massive Machine-type Communication

The second use case is Massive Machine-type Communication, millions of "things" communicating over cellular, a large part of the "Internet of Things" (IoT) story.

In smart city projects, traffic counters and traffic lights communicate with a central control system to better coordinate traffic flow. In more humble circumstances, the tracker inside a FedEx package or on the bottom of a barrel of methylamine might wake up and transmit its GPS coordinates to a central tracking system every hour.

In an example of comparing what is in the document vs. what needs to be done, a new device class 0 was added in a follow-on 3GPP release. This class of device has basic communication functionality and a sleep mode, allowing very low power consumption.

This would allow the package tracker with its built-in cellular radio to be powered from a coin battery with a reliable life measured in years.

20.4.3 Ultra-Reliable, Low-Latency Communications

The third use case is Ultra-Reliable, Low-Latency Communications. Latency is used to mean the time delay caused by the network, and is usually measured as a round-trip. Ultra-Reliable means error correction provided as a network service, built in to the low-level communication protocols.

One application is remote medicine; for example, a surgeon guiding a robot 800 miles away performing heart surgery. Using a wireless link for the communications lowers the cost of installation of the robots, compared to installing a landline service for each machine.

A more lucrative application for this is Augmented Reality (AR) and Virtual Reality (VR), where latency causes lag, and can make users seasick and vomit when their eyes and ears send conflicting signals.

Low latency means being able to communicate rapidly changing sensory input from a server in close enough to real time that the brain synthesizes it into a real-seeming experience.

21 Spectrum-Sharing Roundup: FDMA, TDMA, CDMA, OFDMA

We conclude this course with a recap of spectrum-sharing strategies and generations.

Cellphones transmit and receive signals over shared radio bands, shared spectrum. To separate users so that they do not interfere with one another, service providers use spectrum-sharing methods: Frequency-Division Multiple Access (FDMA), Time-Division Multiple Access (TDMA), Code-Division Multiple Access (CDMA) and Orthogonal Frequency-Division Multiplexing Multiple Access (OFDMA).

21.1 FDMA

Frequency-Division Multiple Access (FDMA) is a spectrum-sharing method where a block of spectrum is divided into small frequency bands called channels. Users are assigned transmit and receive channels.

To communicate voice, the frequency of a carrier centered in a channel is varied or modulated continuously in proportion to the voltage coming out of the microphone – which is an analog of the strength of the sound pressure waves coming out of the talker's mouth.

This is referred to as *analog* radio. The same idea is used for FM radio. It was used in first-generation systems, including AMPS in North America, NMT, and TACS used in various countries.

✓ *FDMA: Frequency Division*
  - *Many radio channels, analog FM*
  - *User assigned one channel*
  - *AMPS (1G)*

✓ *TDMA: Time and Frequency Division*
  - *Digitized voice or data in time slots*
  - *GSM, IS-136 TDMA (2G)*

✓ *CDMA: Code Division*
  - *Transmit codes to represent 1s, 0s*
  - *Same time, same frequency band*
  - *"Spread spectrum"*
  - *IS-95 (2G), 1X, UMTS, HSPA (3G)*

✓ *OFDMA: Orthogonal Frequency Division*
  - *Many radio channels, QAM-64 modems*
  - *RRC dynamically assigns multiple channels*
  - *LTE, LTE-Advanced (4G)*
  - *New Radio (5G)*

FIGURE 34 SPECTRUM-SHARING ROUNDUP

21.2 TDMA

Time-Division Multiple Access (TDMA) means a radio channel shared in time between several users. Users transmit and receive modem signals one after another in a strict order in time slots on a radio channel.

TDMA is called digital, since the modems integrated in the handset and base station move 1s and 0s, which could be digitized speech, text messages, web pages, video, or anything else.

In North America, the IS-136 standard was deployed by some service providers for 2G. This was sometimes called D-AMPS as it was digital implemented on 30 kHz AMPS channels.

In the rest of the world, another form of TDMA called the Global System for Mobile Communications (GSM) was deployed for 2G. GSM became the most popular technology.

21.3 CDMA

Code-Division Multiple Access (CDMA) is completely different. The allocated spectrum is organized into 1.25 or 5 MHz frequency bands called carriers.

Users transmit at the same time, in the same cell, on the same carrier. Each user has a code, which is a binary number 64 bits long. If the user wants to transmit a "0," they send their code. If they want to transmit a "1," they send the mathematical complement of their code.

The codes are such that when codes from multiple users are received, added together, mathematical operations reveal which codes were transmitted.

Sending a 64-bit-long code instead of a single 1 or 0, has the effect of transmitting at a higher bit rate, which spreads the energy of the transmission across a wider frequency channel than non-coded transmission does, and so CDMA is also referred to as "spread spectrum".

CDMA was deployed for 2G by some North American service providers. All service providers deployed CDMA for 3G, including the 1X version, UMTS and the HSPA "data" version.

21.4 OFDM and OFDMA

Orthogonal Frequency-Division Multiplexing (OFDM) is the spectrum-sharing method for the downlink for fourth generation LTE and LTE-Advanced, and for the fifth generation called 5G New Radio.

OFDM is a frequency-division technique with significant improvements on 1G analog FDMA.

In an OFDM system, hundreds or thousands of subcarriers are defined within the main carrier. A modem representing 1s and 0s operates on each subcarrier.

A Radio Resource Controller implements Orthogonal Frequency-Division Multiplexing Multiple Access (OFDMA), dynamically assigning one or more subcarriers to a particular user.

This allows multiple simultaneous users on different subcarriers, as well as communication of high bit rates by splitting the bit stream into multiple parallel streams.

Other aspects of OFDM – of interest mainly to mathematicians and Engineers – include the fact that the combined output waveform is calculated in a single step, and transmitted at the same rate as the subcarrier spacing.

The uplink adds a Discrete Fourier Transform to the processing so that the frequency components of the modem waveform are transmitted instead of the waveform itself. This is called DFT-spread OFDM or Single-Carrier FDMA.

LTE, 3GPP Release 8 is referred to as 4G by most. LTE-Advanced, 3GPP Release 10, sporting 1 Gb/s downloads is the technology called 4G by standards committees.

5G, which is called *New Radio* in the standards committees, also uses OFDM and OFDMA.

Course 2233

Fixed Wireless

1 Infrared

21.5 IrDA

IrDA is a standard defined by the Infrared Data Association (IrDA) consortium. It specifies a way to wirelessly communicate data by turning an infrared light on and off. The IrDA specifications include standards for both the physical devices and the protocols they use to communicate with each other.

A revision specifying 1 Gb/s was adopted in 2009 and plans for 5 and 10 Gb/s standards have been announced.

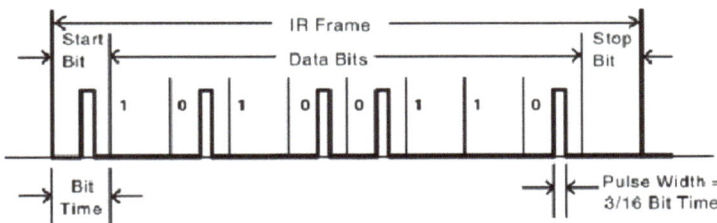

FIGURE 35 INFRARED USING START/STOP/PARITY FORMAT

21.6 Wavelength

IrDA devices communicate using infrared LEDs. The wavelength used is 875 nm, similar to the laser wavelength used on low-cost multimode fiber systems, not visible to the human eye.

IrDA devices conforming to standards IrDA 1.0 and 1.1 work over distances up to 1 m with Bit Error Rate (BER) 10^{-9} and maximum level of surrounding illumination 10 kLux (daylight).

Values are defined for up to 15-degree angle between the receiver and the transmitter. Output power for individual optical components is measured at up to 30 degrees. Directional transmitters (IR LEDs) for higher distances exist, but they don't comply with the required 30-degree angle.

21.7 Start/Stop/Parity

Pulses of light of a duration of 3/16 of the length of the bit interval are used.

The data format is the same as for a PC's serial port, sometimes called *asynchronous, serial* or more accurately, start-stop-parity.

Bit rates for IrDA in its first version ranged from 2.4 to 115 kb/s. Subsequent revisions specified 500 kb/s then 1, 4, 16 and 100 Mb/s with a range of 3 m.

22 Bluetooth

Bluetooth is a set of standards for short-range digital radio communication published by a consortium of companies called the Special Interest Group.

It was originally developed as a wireless link to replace cables connecting computers and communications equipment, operating in the 2.4 GHz unlicensed band.

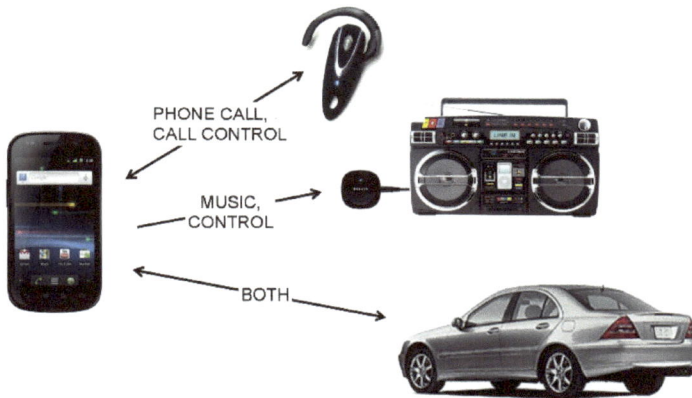

PHONE CALL,
CALL CONTROL

MUSIC,
CONTROL

BOTH

FIGURE 36 TYPICAL USES FOR BLUETOOTH

In practice, Bluetooth is used point-to-point with ten meter range.

In the standard, Bluetooth connections are called piconets, and Personal Area Networks since, in theory up to eight devices can communicate on a channel within a range of 1 to 100 meters depending on the power.

22.1 Data Rates and Variations

The first data rate was 1 Mb/s (0.7 Mb/s in practice), followed by an enhancement to "3" Mb/s (2.1 Mb/s in practice).

A High-Speed variation employs collocated Wi-Fi for short high-bitrate transmissions at 24 Mb/s.

The Smart or Low Energy variation allows coin-sized batteries on devices like heart-rate monitors.

22.2 Frequency-Hopping Spread Spectrum

Bluetooth implements frequency-hopping, where the devices communicate at one of 79 carriers spaced at 1 MHz for 625 microseconds (µs), then hop to a different carrier for 625 µs, then to another, in a repeating pattern known to both devices.

A hop sequence is called a *channel,* and is identified by an *access code.* This is called Frequency-Hopping Spread Spectrum (FHSS), since hopping between 79 carrier's spreads energy across spectrum 79 times wider than one carrier.

It has the advantage of reduced sensitivity to noise or fading at any carrier. If different pairs of devices are using different hop sequences, they can communicate at the same time in the same place without interfering. There are security advantages if the hop sequence cannot be determined by a third party.

22.3 Masters and Slaves, Ticks and Slots

The initiator of communications is called the *master.* It determines the frequency hopping pattern, when the pattern begins, when a packet begins, and when a bit begins.

The packet and bit timing are based on the master's clock, which ticks every 312.5 microseconds. Two ticks make a slot. A slot corresponds to a hop.

The master transmits and the slave listens in even-numbered slots; vice-versa in odd-numbered slots.

To establish the channel, the master derives a channel access code from its Bluetooth address, and indicates the code to the slave at the beginning of every packet. Both master and slave use this to determine the actual frequency-hopping sequence.

Data is organized into Bluetooth packets for transmission… which would more accurately be called frames, as they are broadcast between the devices, not routed. These frames can carry IP packets or other data formats. Bluetooth frames can be 1, 3 or 5 slots long. A bit rate of 2 Mb/s would mean Bluetooth frames are about 150, 450 or 750 bytes long.

22.4 Discovery and Connection

Discovering other devices means sending requests in packets on pre-defined channels called *inquiry scan channels*.

Making a device *discoverable* means it listens on the inquiry channels, and responds to inquiries with information like its Bluetooth address, name and capabilities. This results in a list of Bluetooth devices displayed on the discovering device, such as a smartphone.

Connecting to a device means *paging* the device on its *paging channel*, a channel with access code derived from the target's Bluetooth address.

Devices listen on their paging channel, and respond to pages to establish a session.

Once the session setup protocol is completed on the paging channel, the devices begin communicating on the channel defined by the master.

The frequency hopping pattern can be adapted to skip carriers where the signal to noise ratio is permanently low, to improve overall performance.

Applications include wireless keyboard, mouse and modem connections… though today, 2 Mb/s Bluetooth is likely slower than the modem.

Bluetooth is used to replace wires connecting a phone to an earpiece, or to an automobile sound system for hands-free phone calls while driving. In this case, both two-way audio and two-way control messages are transmitted.

Bluetooth is also used to stream music from a smartphone to a receiver connected to an amplifier and speakers in an automobile or in a living room.

In the future, wireless collection of readings from devices like heart-rate monitors will be widespread.

Each of these types of applications corresponds to a Bluetooth *profile*, which is a specified set of capabilities and protocols the devices must support.

23 Wi-Fi: Wireless LANs

Perhaps the most widespread broadband wireless data communication technology today is wireless LAN technology, also referred to as 802.11, Wi-Fi and hotspots.

This is essentially Ethernet LANs using space as the physical medium instead of copper or fiber, and operating in unlicensed bands at data rates measured in the tens and hundreds of Mb/s, with ranges measured in the tens or hundreds of feet.

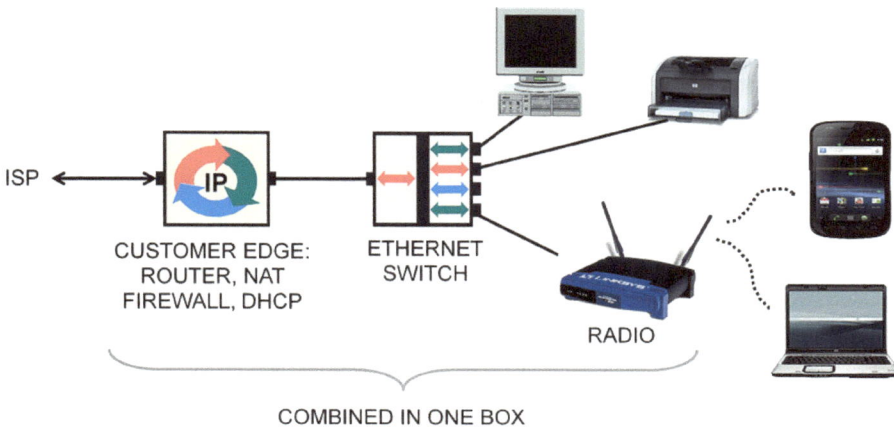

FIGURE 37 WI-FI ACCESS POINT SETUP

23.1 Access Point and SSID

A typical set-up employs a radio base station, called an Access Point (AP), and wireless interfaces built into devices including computers, printers, cameras, phones and music players. Repeaters called range extenders or pods can connect to the AP, by wire or wireless, to provide coverage in an extended area.

The Access Point is usually part of a device called the Customer Edge (CE). The CE also includes an Ethernet switch, routing, Dynamic Host Configuration Protocol (DHCP), Network Address Translator (NAT), firewall and port forwarding, and if supplied by an Internet Service Provider (ISP), will also include a DSL or Cable modem, or for the lucky few, a fiber port.

Those topics are covered in other courses. This lesson concentrates on the radio aspect of Wi-Fi.

The AP defines a Service Set ID (SSID) text string, which it may or may not broadcast. This shows up on the list of available networks on a wireless device trying to connect.

23.2 Half-Duplex

The first five generations of Wi-Fi are half-duplex: only one device can transmit at a time. This means that during file transfers using TCP, the transmitter and receiver must alternate sending data and sending acknowledgements.

Devices transmit only if they have data to send. Otherwise, another device can use the bandwidth. This technique is called statistical Time-Division Duplexing (TDD). It is more efficient than Bluetooth, which reserves half of the time slots for each direction whether they will be used or not.

But it also means that the actual throughput will be less than half the advertised rate during two-way communications, and worse with each added station using the AP.

23.3 802.11 Standards

There are currently six main standards for this technology, published by the 802.11 working group of the Institute of Electrical and Electronic Engineers (IEEE): 802.11a, 802.11b, 802.11g, and the more recent 802.11n, 802.11ac and 802.11ax.

All of these standards operate in unlicensed radio bands, also called Industrial, Scientific and Medical (ISM) bands in North America. "Unlicensed" means that it is not necessary to obtain a license from the national government to emit electromagnetic energy at these frequencies. In other bands it is necessary to obtain a license, which usually involves proving that you will not interfere with anyone else at the requested frequencies in a specific geographic area.

23.3.1 2.4 GHz Band

802.11b and g operate in the 2.4 GHz ISM band, offering a maximum of 11 Mb/s and 54 Mb/s respectively.

802.11b uses Direct Sequence Spread Spectrum like cellular CDMA, and 802.11g implements OFDM like LTE, both allowing multiple Wi-Fi networks at the same time in the same space.

However, many other devices including cordless phones, baby monitors, and Bluetooth operate in this band, meaning significant performance-lowering interference. Microwave ovens operate at 2.4 GHz and can cause radical interference with Wi-Fi communications.

Placing an 802.11 access point on top of a microwave oven results in no wireless LAN while the oven is on because of its interference. It is also possible to listen to 802.11b transmissions on an analog 2.4 GHz cordless phone.

23.3.2 5 GHz Band

802.11a operates in a 5 GHz unlicensed radio band, supporting a maximum of 54 Mb/s using OFDM. The 5 GHz band is relatively free of interference, but the higher frequency also means shorter range and poorer penetration through walls. In practice, line-of-sight between the access point and the terminal are necessary to achieve 54 Mb/s.

802.11n uses the 2.4 and/or 5 GHz bands, optimizing for power to noise ratio between the bands. 802.11n also supports 20 MHz and/or 40 MHz channels – using more of the wireless spectrum when available to enhance performance, and allows parallel transmission using 1 to 4 radios in Multiple-Input, Multiple-Output (MIMO) systems to achieve very high data rates.

In theory, 802.11n will implement 150 Mb/s with a single antenna... but that would be on the moon, where there are no atoms between the transmitter and receiver and no interference. As soon as there is anything between the transmitter and receiver – like water molecules, plaster, concrete and so forth, and/or interference, the power-to-noise ratio and thus bit rate drops.

802.11ac operates at 5 GHz with wider bands and 4x4 MIMO to achieve in practice 500 Mb/s and in theory 3.5 Gb/s.

802.11ax uses up to 160 MHz channels, QAM-1024 and 8x8 MIMO to achieve in theory up to 9.6 Gb/s, OFDMA to share bandwidth efficiently between many users, and device sleep/wake control like LTE and 5G. The 6 GHz unlicensed band is used by 802.11ax.

23.4 Application

Wi-Fi is not used for the "last mile", but rather the "last 10 yards" of a connection to the Internet. It is a short-range access technology from the AP to users. The AP is typically connected to an ISP with wires or fiber.

Wi-Fi is also used to connect printers, security cameras and the like in a LAN with wired and wireless users. The users in this local area can then, for example, print, watch the video feed, and run servers that store and retrieve video.

The LAN also includes the Customer Edge device, which (in Windows-speak) is the default gateway to the Internet.

Voice over IP (VoIP) telephone calls over Wi-Fi is a growth area. A cellphone, iPod, or tablet with 802.11 and Skype minutes to connect over the Internet to a PSTN phone number has no need to pay for cellular service to make a phone call whenever the device has a Wi-Fi Internet connection... which may be "most of the time" in the future.

Proprietary applications like Whatsapp allow communications between Whatsapp users over Wi-Fi access to the Internet.

24 Wi-Fi Security and WPA2

A major concern with wireless LANs is security.

24.1 Address Filtering

Address filtering is one network security measure that can be implemented: setting the access point to only accept connections from specific wireless LAN interfaces, e.g., a laptop.

This is implemented with rules that permit communications to specified Media Access Control (MAC) addresses, and deny communications to any others. The MAC address is hard-coded in each wireless device.

This restricts communications only to wireless devices on the list, protecting against access by unauthorized users – but does not protect legitimate users' transmissions from eavesdropping.

24.2 Eavesdropping

If someone can get physically close enough to receive signals, there is no way to prevent them from eavesdropping on communications, which can include intercepting and re-using usernames and passwords and intercepting and "wiki-leaking" sensitive information.

This is particularly troublesome in coffee shops, airports and anywhere else the communications are not encrypted, "open" hotspots.

In 2010, a plugin for the Firefox browser was made available that allowed someone sitting in such a coffee shop to eavesdrop on everyone else's communications. If a person transmits their username and password in the clear over Wi-Fi, the eavesdropper can with one click of the mouse re-use the username and password to log into that person's account. This means that secure encryption of communications over the airlink is mandatory.

24.3 Airlink Encryption

If it can be ensured that the users always, without fail, implement client-server encryption (sometimes called Transport Layer Security... though anyone who has taken the OSI Layers course will know it is presentation layer security), by using VPN software for connecting to work, typing https:// for all web surfing, and using encrypted email communications, then there is no need for encryption of the airlink.

However, users can not be relied upon, so encryption of the communications on the airlink between the access point and terminal must be implemented whenever possible.

This is implemented by entering a password and the generation and sharing of encryption keys.

The terminal encrypts its data and the base station decrypts with these keys and vice-versa.

24.4 WEP and WPA2

Wired Equivalent Privacy (WEP) was the first encryption algorithm for wireless LANs; but its use is not recommended as there are tools available that can determine the key in a matter of minutes.

Wi-Fi Protected Access 2 (WPA2) with its AES encryption should be implemented when possible. WPA2 is the Wi-Fi Alliance's name for a security amendment to 802.11 initially referred to as 802.11i then incorporated into the 802.11 standard in 2007.

WPA2 uses a framework called 802.11X Port-based Network Access Control to exchange security messages and block or unblock Internet traffic for a device. 802.11X was initially designed to block or unblock physical hardware ports on LAN switches.

For wireless LANs, the term "port" is more of a notion, identifying traffic for a device, and the "switch" is the Wi-Fi AP.

As illustrated in Figure 38, two types of ports are defined on the AP.

The Uncontrolled Port is used by all devices to transmit and receive only security messages conforming to the Extensible Authentication Protocol (EAP) for communication of passwords and encryption keys.

The device desiring to communicate is called the supplicant, and the function in the AP it communicates with is the authenticator. The messages are defined in EAP "methods," including EAP-PWD for passwords and EAP-IKEv2 for key exchange.

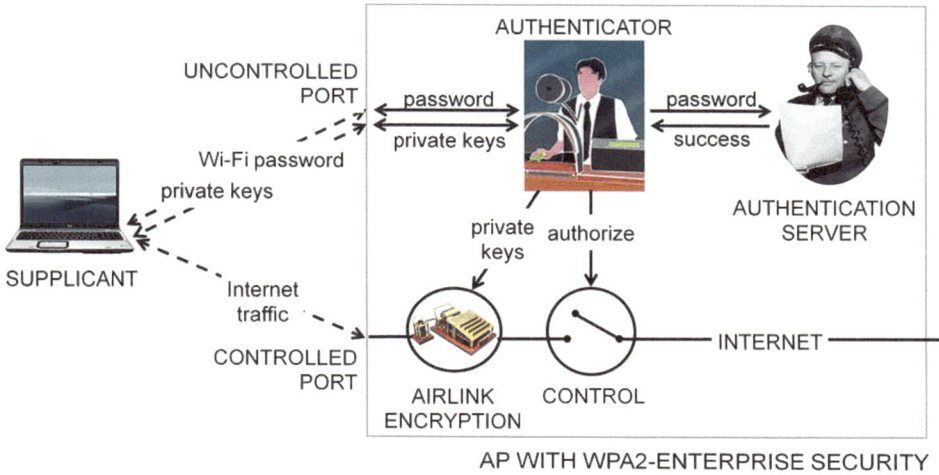

FIGURE 38 WPA2-ENTERPRISE HAS MULTIPLE PASSWORDS

Each device is associated with its own Controlled Port, through which application traffic to other devices flows. In most cases this is traffic to and from the Internet. Sometimes, the other device is locally-attached; for example, the HTTP server front-ending the AP's control panel.

24.5 WPA2 Enterprise vs. Personal

The full implementation, called WPA2-Enterprise, supports multiple users with different passwords. The authenticator communicates with an authentication server using messages conforming to the RADIUS protocol to validate the supplied password, and on success, negotiates a 256-bit private key called the Pairwise Master Key (PMK) with the supplicant.

A simpler implementation, WPA2-Personal, supports only one password, which either is the 256-bit PMK or a passphrase of 8 to 63 characters that is used to generate the PMK. This is referred to as a Pre-Shared Key (PSK). Since there is only one password, an authentication server is not required.

WPA3 is an incremental improvement with security enhancements.

24.6 Operation

After authentication, private keys for the session are generated and the controlled port is authorized. The PMK and other information is used to generate a private key called the Pairwise Transient Key for bulk encryption of normal traffic between the AP and the device, and a Group Temporal Key for multicast traffic between the AP and multiple devices. The keys are exchanged in EAP messages.

The controlled port is set to unauthorized when the authentication process begins, meaning that no application traffic is permitted through the AP.

After successful authentication and key exchange, the controlled port is authorized, allowing Internet and other application traffic through the AP. This traffic is encrypted on the airlink, that is, between the AP and the device, using strong AES-based encryption.

25 Point-to-Point Microwave

Point-to-point microwave systems are used in two main applications:

*   By phone companies to implement communications where it is not practical to pull a fiber, such as over mountain ranges and rivers, and

*   By all kinds of users to implement high-capacity short- or medium-range point-to-point communications without paying the phone company for a circuit.

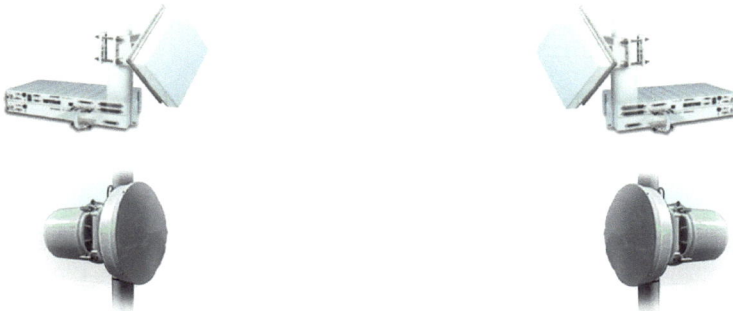

FIGURE 39 POINT-TO-POINT MICROWAVE ANTENNAS

25.1 Frequency Bands

In addition to the unlicensed bands at 2.4 and 5.8 GHz, microwave systems also operate on the following frequency pairs (transmit and receive FDD), which require a license from the FCC in the US or the CRTC in Canada:

5925-6875 MHz

6245-6930 MHz

7125-7725 MHz

7725-8275 MHz

8275-8750 MHz

10550-10680 MHz

10700-11700 MHz

12750-13250 MHz

14400-15350 MHz

17700-19700 MHz

38600-40000 MHz

Rates from 1 DS1 to 3 DS3 with ranges up to 50 miles are available in off-the-shelf systems. As usual, this depends on the power, obstructions, interference and other factors. Harris is one of the leading manufacturers of these systems.

25.2 Fading, Diversity and Error Correction

Microwave radio suffers from random fading due to destructive interference of signals received via different bounce paths – which can change from one instant to the next.

To ensure communications, *diversity* is required. Diversity means a second communication system.

Frequency diversity is a second system at a different carrier, or a spread-spectrum strategy. Space diversity is a parallel system operating at the same frequency a short distance away. Time diversity is transmitting the signal more than once.

Wireless communications, including point-to-point microwave, often incorporates error-correcting codes, creating longer codes than necessary and only using some of them, evenly spaced, so that errors can be corrected by the receiver by choosing the closest valid code at the receiver.

This is also called Forward Error Correction (FEC) or redundant coding.

26 3.5 GHz Broadband Fixed Wireless Internet

When the density of homes is above 500 homes per square kilometer (5 homes per hectare = about 2 homes per acre), the economics of installing fiber infrastructure with a long-term payoff becomes interesting.

Broadband to the home in areas without sufficient density of paying customers, and without suitable existing cabling infrastructure is usually most cost-effectively delivered wirelessly. This is called *fixed wireless*, or wireless home Internet.

Wireless broadband to the home means a wireless link between an antenna on a tower and an antenna fixed to the outside of the home, which is connected (and powered) by a cable to an indoor device, which provides Wi-Fi and physical LAN jacks in the home.

FIGURE 40 BROADBAND FIXED WIRELESS ANTENNA

26.1 Initial Deployment

For cost reasons, it would be ideal to provide both mobile Internet to cellphone users, and fixed wireless for broadband home Internet from a single radio technology and a single set of towers and base stations.

At present, the capacity of mobile LTE networks is such that download speeds fall far below "broadband" on weekend evenings when many people are watching video.

Due to licensing, spectrum cost and regulations, the initial rollout of fixed wireless broadband Internet to the home is as a standalone sectorized radio system.

While the antennas would be typically on the same towers as the mobile network antennas, the fixed wireless system is not accessible by cellphones.

The 3.5 GHz band (3350 MHz to 3700 MHz) allows the allotment of numerous channels 10 to 40 MHz wide. MIMO is used to increase the maximum bitrate available to users.

A 40 MHz channel yields 200 Mb/s downloading given a spectral efficiency of about 5 bits/second per Hertz for LTE.

In the initial rollout, LTE and its Radio Resource Controller is used to share the capacity between subscribers in a given sector.

Offering 25/1 service allows the carrier to support between 100 to 150 users per sector, using 10 Gb/s backhaul for each multi-sector base station.

26.2 Near-Field Interference

While the 3.5 GHz band supports high capacity airlinks, it requires line of sight between the antennas, and is significantly affected by near-field interference, for example trees growing in front of a residence.

This may require the antenna to be mounted on a mast on the roof of a residence, on a 15-meter tower, or mounted remotely with a backhaul to the residence.

26.3 Subsequent Deployments

Broadband wireless will be deployed on additional spectrum including the 3.7 GHz band (3700 – 4000 MHz) with 20 to 100 MHz wide channels, and the 2.5 GHz band with 50 MHz-wide channels.

All new builds will employ 5G technology to share the capacity between users in a neighborhood. With its massive MIMO and enhanced transmission modes, 5G New Radio is 40% more efficient than LTE.

Along with more sophisticated customer antennas, this will in the future enable 50/10 as a standard wireless broadband home Internet service, particularly for areas not served by DSL, cable or fiber to the home.

27 Low-Power Wide-Area (LPWA) Radio Networks for IoT

Low-Power Wide-Area (LPWA) systems, also called Low-Power Wide-Area Networks (LPWANs) are wireless systems for sensor data reporting and remote control.

There are four main technologies, divided into two groups: technologies deployed by mobile (cellular) carriers, and technologies deployed by non-cellular carriers.

FIGURE 41 LPWA TECHNOLOGIES

The technologies deployed by cellular operators are Narrowband Internet of Things (NB-IoT) and Long Term Evolution category M1 (LTE-M).

The technologies deployed by non-cellular operators are sigfox and LoRa (short for Long Range).

27.1 Technologies Deployed by Non-Mobile Network Operators

Sigfox and LoRa operate in unlicensed spectrum often called Industrial, Scientific and Medical (ISM) bands, at relatively low frequencies. In Europe the 868-MHz band is used; in the US it is 915 MHz; and 923 or 433 MHz in Asia.

Sigfox is a French company that established a base of customers with water meters from the French water authority Veolia and has expanded internationally. It provides the leanest communication service, using 100 Hz of bandwidth to move tiny data packets with payloads of 12 bytes upload and 8 bytes download, with limits on the number of messages per day. The theoretical data rate is 600 bits per second. This reduces the subscription costs and extends battery life. It can support 1 million devices per base station transceiver.

LoRa is a competing technology also using ISM bands, using 125 kHz of bandwidth to achieve a theoretical 50 kb/s and 40,000 devices per base station. LoRa employs spread-spectrum coding, meaning the modem signal is spread over a wider frequency band than normal. This allows better performance in the presence of noise or jamming. The LoRa Alliance was created to foster interoperability between devices.

27.2 Technologies Deployed by Mobile Network Operators

Carriers offer NB-IoT and LTE-M services in licensed frequency bands.

NB-IoT operates on 200 kHz channels, often in the "guard bands", i.e., unused spectrum between bands used for conventional cellular, supporting around 200 kb/s data rates depending on carrier implementation. It is designed for fixed, ultra-narrow-bandwidth IoT applications. It purports to provide better connectivity in subterranean locations such as basements, utility vaults and sensors located deep within buildings. It does not support mobility, i.e., maintaining connection as the sensor moves out of range of a base station.

LTE-M operates in approximately 1 MHz bands on licensed spectrum, supporting up to 1 Mb/s data rates (depending on carrier implementation), as well as voice and mobility. As the name would suggest, it is intended for existing LTE cellular networks, to provide extended coverage to IoT applications. Due to its lower latency and higher bandwidth than NB-IoT, LTE-M is suited for IoT applications where devices are in motion and real-time data is required.

The sigfox and LoRa systems gained an early market share. Carrier NB-IoT and LTE-M systems are expected to significantly outpace sigfox and LoRa in terms of connected devices going forward.

27.3 The 5G Steamroller

In the long run, support for IoT using 5G deployed by Mobile Network Operators will render most of these systems obsolete.

28 Satellite Communications

28.1 Introduction

In this last lesson of the course, we will take a quick overview of communication satellites, understanding the basic principles and the advantages and disadvantages of the two main strategies: Geosynchronous Earth Orbit and Low Earth Orbit.

Communication satellites are orbital platforms that carry multiple base station transceivers with antennas pointed towards the surface of the earth. Instead of base station transceiver, the term transmitter/responder or *transponder* is used in the satellite business.

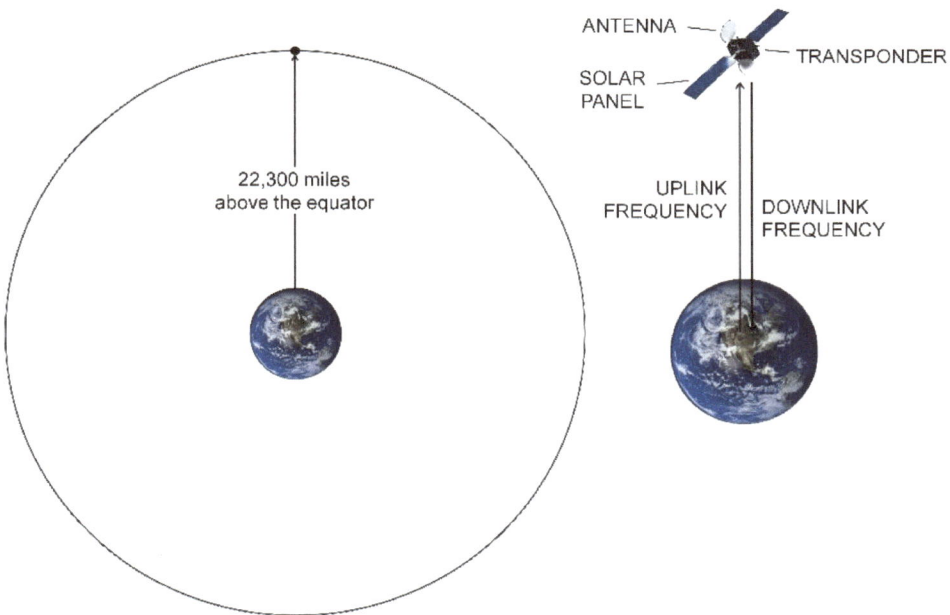

FIGURE 42 GEOSYNCHRONOUS SATELLITES

In two-way systems, radio signals are transmitted from the earth to the transponder, which responds with radio signals directed back down to the surface at a different frequency to avoid interfering with the surface transmitter.

There are two basic choices for the orbits of communication satellites: geosynchronous orbit or low earth orbit.

28.2 Geosynchronous Earth Orbit

Geosynchronous satellites are parked 22,300 miles (35,680 km) above the surface of the earth above the equator (Figure 42). At that radius, the orbital speed is the same as the rotational speed of the earth, and hence the satellite appears to stay in the same spot in the sky. This is called geosynchronous or geostationary, depending on your point of view.

Geosynchronous communication satellites are operated by the International Telecommunications Satellite Organization (Intelsat), the International Marine Satellite Organization (Inmarsat), numerous private companies, government and military.

Each country has the right to spots in geosynchronous earth orbit above their country based on international agreements. The countries can use these spots, or lease them. Occasionally, there are disagreements as to who has the right to a spot and two countries will launch a satellite and park it in close proximity to the other, causing interference.

28.2.1 Path Delay

The three main advantages of geosynchronous satellites are broadcast, broadcast and broadcast: a transponder on a geosynchronous platform is 22,300 miles up in the sky - and if desired, can provide radio coverage to a third of the earth's surface.

The main disadvantage is path delay: the radio waves have to travel 22,300 miles up and 22,300 miles down. At the speed of light, that takes about 1/4 second each way.

For interactive communications between two points on the earth via a geosynchronous satellite, that means up and down for the inquiry, and up and down for the response, a total of just under one second delay. If the two locations cannot see the same satellite, then an intermediate ground station must be used, meaning a path delay of about two seconds.

This wreaks havoc with the protocol people use to decide who gets to talk next during a phone call, and the extended delay causes users to hear echoes that are normally suppressed.

No-one likes using use geosynchronous satellites for phone calls; trans-oceanic fiber optic cables are preferred because they are much shorter, meaning that the path delay is negligible; about 25 milliseconds from New York to Paris on fiber, for example.

One-way communications is the natural application for geosynchronous satellites. Television is the biggest market. Radio-frequency modems communicate video that has been digitized, coded using MPEG-2 or H.264 (MPEG 4 Part 10) and encrypted from a Digital Broadcast Center of a satellite TV company up to the transponder, which repeats the modem signal back to the earth.

Access to the Internet is also implemented on geosynchronous satellites for those who do not have DSL, Cable, cellular or fiber terrestrial links available. The "upload" path from the customer to the Internet can be modems over a regular phone line, and the download path is via satellite. This makes the customer premise electronics cheaper and cuts the delay in half. Two-way satellite communications is also available.

Another service based on geosynchronous satellites is *Very Small Aperture Terminal* (VSAT), which means "small dishes" in plain English. An application for this two-way wide-area data communication service is government and military communications, and communications for remote areas.

28.3 Low Earth Orbit

The path delay problem of geosynchronous platforms can be fixed by bringing the satellite closer in. These type of communication satellites are called Low Earth Orbit (LEO), closer to the earth as illustrated in Figure 43. These can be used for voice communications because the path delay is reduced to an acceptable level.

This introduces two different problems: the satellites do not stay in the same position in the sky to an earthbound observer, and the coverage or *footprint* of the satellite is reduced.

Multiple satellites to ensure coverage and a switching system for doing handoffs from one satellite to another as they move are required. The handoff problem is similar to cellular radio; except that in cellular, the base stations are stationary and the users move around, whereas with LEO satellites, the users are more or less stationary and the base stations move around.

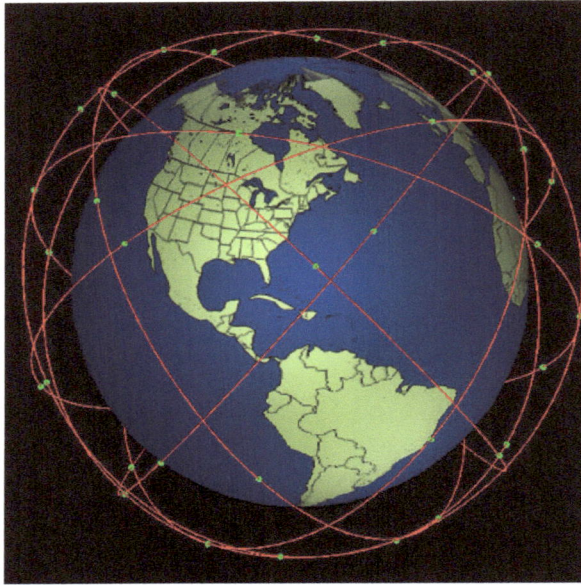

FIGURE 43 LOW EARTH ORBIT SATELLITES

Motorola's Iridium project was one example of this. They planned to launch 77 satellites (Iridium is element number 77 in the periodic table of the elements), but went live after deploying only 66 satellites.

Gaps in coverage, poor in-building penetration and difficult data communication over the analog radio system led to poor user response and Iridium was a financial failure.

Motorola announced that they would "de-orbit" the satellites at a loss of five billion dollars. At the last minute, an entrepreneur with some contracts with mining firms purchased Iridium for 25 million dollars.

In 2017 and 2018, the SpaceX Falcon 9 rocket was used to launch 75 of 81 Iridium 2 satellites to form Iridium Next.

The original Iridium satellites were de-orbited in 2019.

Running on Iridium Next, Iridium OpenPort Internet Service is available at rates of 128 to 512 kb/s. Iridium L-Band broadband service is up to 1.5 Mb/s down, 0.5 Mb/s up.

Iridium Certus is the branding for a new set of features enabled by Iridium Next: mobile satellite communications across maritime, IoT, aviation, land mobile, and government applications.

Among other services hosted by Iridium Next is the Aireon aircraft tracking and surveillance system. This system will provide air traffic control organizations and aircraft operators that purchase the service with real-time global visibility of their aircraft. The search for flight MH370 that disappeared in 2014 would have been easier with such satellite tracking.

Other LEO companies include Orbcomm, Globalstar and Starlink. Orbcomm was a joint venture between Teleglobe and Orbital Sciences Corporation, intended to provide two-way data communications and the capability to track trucks on highways and tanks on battlefields.

Globalstar is a consortium of telecommunications companies operating a constellation of 48 low earth orbit satellites, and acting as a wholesale provider of mobile and fixed satellite-based telecom services.

Globalstar's business plan emphasized the integration of local wireless and satellite communications: one phone could be used to communicate virtually anywhere on the planet. Globalstar transmits calls from a wireless phone or fixed phone station to a terrestrial gateway, where they are passed on to existing fixed and cellular telephone networks in more than 100 countries on 6 continents.

28.3.1 Starlink

The SpaceX Starlink System is being launched 60 satellites per Falcon 9 flight, slated to finish with 12,000 satellites by the mid-2020s.

The satellites are positioned at three orbits: 340 km (210 mi), 550 km (340 mi), and 1100 to 1325 km (690 - 830 mi).

Starlink intends to provide broadband to the masses at prices competitive to terrestrial services.

After a launch in 2020, the chief Engineer at Starlink said "we're still a long way from watching cat videos in 4K, but we're on track to get there".

In 2022, it indeed became possible to watch cat videos and guilty dog videos, with download speeds up to 1 Gb/s in bursts, for US$600 to $800 equipment cost and $110 per month service.

FIGURE 156 STARLINK 2ND GENERATION ANTENNA

Starlink's second generation antenna is a 19 x 12 inch flat-panel antenna that points skyward. There are actually many antennas under the cover; very sophisticated signal processing adjusts the power and phase of each in real time to perform *beamforming*, focusing the power on the satellite transceivers.

Similarly, many Starlink satellites work together to focus power on the antenna as they transit overhead. The time a Starlink satellite is visible to an antenna before it fades into the noise horizon is measured in minutes.

About Teracom

Public Seminars

Instructor-led training is the best you can get, allowing you to ask questions and interact with classmates. Teracom's public seminars are in-person and live online instructor-led courses geared for the non-engineering professional needing a comprehensive overview and update, and those new to the business needing to get up to speed.

Private Onsite and Online Seminars

Since 1992, we have provided high-quality on-site and live online private instructor-led training in telecommunications, data communications, IP, networking, VoIP and wireless at hundreds of organizations ranging from Bell Labs to the Defense Systems Information Agency.

We have built a solid reputation for delivering high-quality training programs that are a resounding success. We would like to do the same for you! Please contact us via teracomtraining.com for more information.

Online Courses and TCO Certifications

Upgrade your knowledge - and your résumé - with high-quality telecom training courses by Teracom coupled with certification from the Telecommunications Certification Organization:

- Certified Telecommunications Network Specialist (CTNS)
- Certified Telecommunications Subject Matter Expert (CTSME)
- Certified Telecommunications Analyst (CTA)
- Certified Wireless Analyst (CWA)
- Certified VoIP Analyst (CVA)

TCO Certification is proof of your knowledge of telecom, datacom and networking fundamentals, jargon, buzzwords, technologies and solutions.

Guaranteed to Pass and repeat courses anytime with the Unlimited Plan!

Join our thousands of satisfied customers including:

the FBI Training Academy, US Marine Corps Communications School, US Army, Navy, Air Force and Coast Guard, the NSA and CIA, DoJ NSD, IRS, FAA, DND, CRA, CRTC, RCMP, banks, power companies, police forces, manufacturers, government, dozens of local and regional phone companies, broadband carriers, individuals, telecom planners and administrators, finance, tax and accounting personnel and many more from hundreds of companies.

Teracom's GSA Contract GS-02F-0053X for supplying this training to the United States is your assurance of approved quality and value.

Visit us at teracomtraining.com to get started today!